Lecture Notes in Mathematics 1664

Springer
Berlin
Heidelberg
New York
Barcelona
Budapest
Hong Kong
London
Milan
Paris
Santa Clara
Singapore
Tokyo

Martin Väth

Ideal Spaces

Author

Martin Väth
Mathematisches Institut
Universität Würzburg
Am Hubland
D-97074 Würzburg, Germany
e-mail: vaeth@cip.mathematik.uni-wuerzburg.de

Cataloging-in-Publication Data applied for

Die Deutsche Bibliothek - CIP-Einheitsaufnahme

Väth, Martin:
Ideal spaces / Martin Väth. - Berlin ; Heidelberg ; New York ; Barcelona ; Budapest ; Hong Kong ; London ; Milan ; Paris ; Santa Clara ; Singapore ; Tokyo : Springer, 1997
(Lecture notes in mathematics ; 1664)
ISBN 3-540-63160-7

Mathematics Subject Classification (1991):
Primary: 46E30, 46E40;
Secondary: 46A45, 46B45, 28A20, 28A35, 28E15, 46B10

ISSN 0075-8434
ISBN 3-540-63160-7 Springer-Verlag Berlin Heidelberg New York

Typesetting: Camera-ready TEX output by the author
SPIN: 10553283 46/3142-543210 - Printed on acid-free paper

Table of Contents

1. Introduction

Ideal spaces are a very general class of normed spaces of measurable functions, which includes Lebesgue, Orlicz, Lorentz, and Marcinkiewicz spaces as well as weighted and combined forms of these spaces. Sometimes these spaces are also called *Banach function spaces* or *(normed) Köthe spaces.*
One intention of this text is to summarize elementary properties of these spaces. In contrast to the existing work on this topic like [48] and [50] we concentrate on the general case of vector-valued functions on arbitrary measure spaces (covering e.g. Lebesgue-Bochner spaces). In the second half we concentrate on spaces with mixed norm and their generalization, and on calculus with vector-valued functions with values in an ideal space. This has applications in the theory of partial integral and partial differential operators. Thus, our results are applicable in particular to integro-differential equations of Barbashin type, which were the main motivation for this text. Another surprising simple consequence of the presented results are some powerful theorems on continuity of Hammerstein operators.

Overview and Summary of Results As we already mentioned, an extension to the summarizing texts [48] and [50] lies in the fact that

1. we usually consider vector-valued functions (especially functions with values in an infinite-dimensional Banach space) and
2. whenever possible, we assume that the domain of definition of these functions is an arbitrary measure space (not just a finite or σ-finite measure space).

The first extension forces us to justify the name of ideal spaces: In the real-valued case these spaces are precisely the L_∞-ideals. So in Sect. 2.1 we consider in which sense this characterization also holds for our spaces. It will turn out that this still is true in almost any case (Theorem 2.1.3). However, we will see that, surprisingly, this is in contrast to the very similar strong characterization property (see Definition 2.1.5).
Here we should emphasize that none of our proofs uses the axiom of choice or the continuum hypothesis, so they are all 'constructive'.

One of the most important questions is of course the connection between convergence in the norm of an ideal space and other notions of convergence like convergence a.e., in measure, or in measure on sets of finite measure (we will generalize and unify these notions to *extended convergence* and prove elementary properties of this new type of convergence in Sect. 2.2).

The first result in this direction is Theorem 2.2.3, which states that it is not possible that a sequence converges in norm to one function, while it converges extended to a different function.
This result is trivial only for finite measure spaces (and thus also for measure spaces with the *finite subset property*), because for finite measure spaces Zabrejko proved in [50] that convergence in norm implies convergence in measure (we will give a shorter proof for this fact and state some of its consequences in Sect. 3.1. However, we will see there that Zabrejko's result can not be extended to infinite measure spaces).
Also the converse of this result is important: Under which conditions does extended convergence imply convergence in norm? Answers to these questions are given in Sect. 3.3, where we prove Vitali's and Lebesgue's convergence theorems for ideal spaces (Theorems 3.3.3 and 3.3.5). In a simpler form these theorems are already known for finite measure spaces. For arbitrary measure spaces we have to use an appropriate definition of *regular* spaces by dividing spaces in an *inner-regular* and an *outer-regular* part, where the outer-regular part covers the 'non-finite' parts of the measure space. For σ-finite measure spaces our notion of regularity is the usual one, and we will see in that section by many examples and elementary properties that in the general case it is the natural one.
In this section we also state a more powerful version of Vitali's theorem in continuous measure spaces (Theorem 3.3.4), which even seems to be unknown for L_p-spaces. Finally in this section we establish a method of estimating $\limsup \|x - x_n\|$ in some ideal spaces, which may be useful for $x \neq 0$, even if $x_n \not\to x$ (Theorem 3.3.6 and its corollaries).

In Sect. 3.2 we establish necessary and sufficient conditions for the completeness of a pre-ideal space. Here again the extension to the vector-valued case and to arbitrary measure spaces forces us to vary the known proofs. The most useful sufficient condition is the following property: The limit function of an almost everywhere convergent norm-bounded sequence belongs to the space. But on arbitrary measure spaces we have the strange phenomenon that we may not assume a priori that this limit function is finite. This is the reason, why in the definition of extended convergence in Sect. 2.2 we also include the case that the limit function is not finite (even for vector-valued functions).
In the real-valued case this property is called *weak Fatou property* in [48]. We call such a space *semi-perfect*. The more restrictive Fatou property is called *perfect* in our text, according to [50], where also the notion *almost perfect* is introduced. We generalize this concept to *almost α-perfect* spaces and justify these definitions by examples.

Furthermore, we give some new proofs of elementary properties of the Lorentz space, which on σ-finite measure spaces might be considered as the perfect hull of a given pre-ideal space.
Finally, in this section we characterize the perfect ideal spaces in terms of the underlying linear metric space of measurable functions (Lemma 3.2.11).

In Sect. 3.4 we will give a proof of Lorentz' important Theorem 3.4.5, which does not refer to the axiom of choice. Some parts of that proof differ from the literature we know. The theorem allows us to characterize semi-perfect, almost α-perfect and α-perfect pre-ideal spaces in terms of the second associate space (or, equivalently, in terms of the Lorentz space) on σ-finite measure spaces. Furthermore, we briefly discuss the non-σ-finite case. Also in this section we consider integral functionals on ideal spaces. Especially, we will prove that for ideal spaces these functionals are automatically bounded (Theorem 3.4.3), and we will characterize them without referring to a supremum (Corollary 3.4.2). Again, this is much harder for the vector-valued case than the well-known results for real-valued functions (in particular, we will make use of the majorising principles proved in Appendix A.2).

In Sect. 3.5 we will discuss the question, in which way the integral functionals are the only bounded functionals on a real-valued pre-ideal space X over some σ-finite measure space. It is already well-known that this is the case for regular X, and, using the axiom of choice, regularity is even necessary. But we will show that in Solovay's model of set theory [44] this is true without the regularity of X. In particular, provided that Solovay's model exists, you can not prove the existence of a non-integral bounded functional on X without applying a strong form of the axiom of choice. The results are applied for some statements about reflexivity.

In the following sections we consider ideal spaces on product measures $T \times S$ (to define this properly, we have to assume that T and S are σ-finite). In Sect. 4.1 we consider the simplest class of such spaces, namely that of product measurable functions x defined in the natural way by the mixed norm $\|x\| = \|t \mapsto \|x(t,\cdot)\|_U\|_V$, where U and V are ideal spaces. The first problem is that this in general does not define a linear space, because it is not clear that $t \mapsto \|x(t,\cdot)\|_U$ is measurable (if defined). A classical theorem of Luxemburg and Gribanov (presented in Sect. A.3) ensures that this is true for perfect ideal spaces U, but we also give some sufficient conditions for almost perfect pre-ideal spaces and for the typical examples of not almost perfect pre-ideal spaces. We also examine other properties of spaces with mixed norms, in particular we determine their support, and check various completeness, perfectness and regularity properties. Furthermore, we gain the formula $[U \to V]' = [U' \to V']$ for the associate space.

The concept of spaces with mixed norm may be generalized by the spaces with mixed family norm, where we replace the space U by a family $U(\cdot)$ of spaces, and where the norm is then defined by $\|x\| = \|t \mapsto \|x(t,\cdot)\|_{U(t)}\|_V$. If possible, we will discuss all problems for spaces with mixed norms also for spaces with mixed family norms in Sect. 4.3. However, in some cases this can not be done. In particular, we will show in an astonishing example that their support is not determined by the support of $U(t)$ and V. Also the Luxemburg-Gribanov theorem has no easy generalization, and we introduce other methods to ensure the measurability of $t \mapsto \|x(t,\cdot)\|_{U(t)}$, which cover the case of $U(t)$ being Orlicz spaces generated by finite Young functions.

While Sect. 4.1 and 4.3 deal with the problem of the construction of spaces over $T \times S$ from spaces over T and S, Sect. 4.2 deals with the opposite problem: Given a function space over $T \times S$, we define function spaces over T and S in a natural way and examine their completeness, perfectness and regularity properties.

Section 4.4 deals with the integration and differentiation of functions with values in ideal spaces. One might suspect that this is equivalent to applying a partial integral or a partial differential operator to a corresponding function on the product measure. We will see that this indeed is the case under natural assumptions. The results in this section generalize the corresponding results for L_p spaces in [8], [15], and [22]. Since the main tool in these works is Fubini's theorem one also might consider some theorems of this section as a generalization of Fubini's theorem for ideal spaces. This explains, why our proofs for ideal spaces need more preparation than in the L_p case. In fact, we use some deeper results of the previous sections.
For integration of functions we find a completely satisfying answer (Theorem 4.4.3). For the natural criterion for measurability we can show necessity in all pre-ideal spaces, but sufficiency only in regular spaces (the reason being that it is not true in general; we give a counterexample). The results about differentiability follow from that about integrability.

It is known that in Solovay's model any linear operator defined on a Banach space is continuous (even more is known, see [17] and [47]). This is our motivation to prove in Sect. 5.1 for a large class of linear operators that they are continuous, if they map an ideal space into some proper space (Definition 5.1.1) – even if you do assume the axiom of choice. This class includes multiplication, integral, and partial integral operators and combinations of those, provided the occurring functions take values in finite dimensional spaces. But the results also cover many (maybe even all) cases, where the functions in consideration take values in infinite-dimensional spaces. The results in this section generalize Banach's classical theorem about continuity of integral operators in L_p spaces and similar results like e.g. in [50].

In Sect. 5.2 we first show that the (nonlinear) superposition operator $Fx(s) = f(s, x(s))$ with a Carathéodory function f is continuous in the interior of its domain of definition, if it acts from an ideal space into a regular ideal space. For x taking values in $\mathbb{R}$ this result is well known (see e.g. [4]), but for vector-valued x (in particular for x taking values in an infinite-dimensional space) we have to develop a completely new method for the proof. We also consider what happens on the border of its set of definition, and prove some sufficient conditions for F to be uniformly continuous on balls. Combining these results with the theorems of the previous sections about spaces with mixed norms and weighted projections of spaces, we find similar results for the Hammerstein operator in Sect. 5.3.

The last Sect. 5.4 gives some examples, how the abstract theory might be applied for some classes of integro-differential equations and partial integral equations.

The sections in the appendix do not deal with ideal spaces, but with measurable functions with values in Banach spaces, especially with operator functions. In Sect. A.1 we are concerned with the measurability of the function $s \mapsto A(s)x(s)$ (with A being an operator function), which is often used without proof. The results in Sect. A.2 are much deeper. Here the problem is, given any measurable linear operator function A and a measurable function y, to find a *measurable* function x, $|x(s)| = 1$, which 'maximizes' A up to the error $|y|$, i.e.

$$|A(s)x(s)| \geq \|A(s)\| - |y(s)|$$

for almost all s, where $\|A(s)\|$ denotes the operator norm of $A(s)$. For σ-finite measure spaces we can even find an essentially countable valued such function x (Theorem A.2.1), but it turns out that the generalization to arbitrary measure spaces is quite hard.
Using the axiom of choice, the result extends to measure spaces with the *direct sum property.* Thus we study this property and find an interesting characterization of this property, which seems to be unknown in literature. We also give a strange example of a space which has the direct sum property, if we assume the continuum hypothesis, but does not have the direct sum property in Solovay's model.
Finally, for the integral version of the above problem, to find a measurable function x, $|x(s)| = 1$, satisfying

$$\int_S |A(s)x(s)|\, ds \geq \int_S \|A(s)\|\, ds,$$

we surprisingly can give an affirmative answer in arbitrary measure spaces S (Theorem A.2.2), even without using the axiom of choice. It is this theorem that we need in Sect. 3.4.

Section A.3 deals with a similar problem as the previous section for the special (functional-valued) function $A(t)u = \int_S |x(t,s)|\, u(s)ds$, where now the measurability of A is replaced by the fact that x is supposed to be product-measurable on $T \times S$ (thus we consider only the σ-finite case). Similar as in Sect. A.2 we ask, whether it is possible to find a product-measurable function u with $\|u(t,\cdot)\| \leq 1$, such that

$$\int_S |x(t,s)|\, |u(t,s)|\, ds \geq \sup_{\|u\|\leq 1} \int_S |x(t,s)|\, |u(s)|\, ds - |y(t)|$$

for a nonvanishing measurable error-function y. An affirmative answer will be given (Theorem A.3.2), using a theorem of Luxemburg and Gribanov (Theorem A.3.1). This theorem and its obvious generalizations are used in Sect. 4.1 and 4.3 to gain results about the measurability of $t \mapsto \|x(t,\cdot)\|_U$, and to prove the nice formula $[U \to V]' = [U' \to V']$.

Acknowledgement. This text is the outcome of two years of research done at the University of Würzburg in the framework of a DFG project (Az. Ap 40/6-2). Financial support by the DFG Bonn is gratefully appreciated.

2. Basic Definitions and Properties

2.1 Ideal Spaces and their Characterization

By a *measure space* (S, Σ, μ) we mean a nonnegative σ-additive set-function μ, defined on a σ-field Σ, whose elements are all subsets of S. We usually just write S for the measure space, and call the elements of Σ measurable sets. Measurable functions are defined as in [15]. Observe that by this definition the usual theorems hold true, even if S is not complete! However, we define null sets as all subsets of sets of measure 0, and write $\text{mes}E = \mu(D)$, if $D \in \Sigma$ and $E = D \cup N$ for some null set, i.e. the terms 'a.e.' and 'mes E' are meant in the sense of the Lebesgue extension of (S, Σ, μ). For the integral we do not mention μ and Σ and just write

$$\int_S x(s)ds.$$

If T and S are σ-finite measure spaces, its product space $T \times S$ is defined as in [15].

Some words about the axioms we are going to use: If not stated explicitly, we will not use the *continuum hypothesis* or the *axiom of choice* (we will just use them in some counterexamples to show that our results may not be sharpened). Instead of the latter we assume its weaker form, the principle of dependent choices, to be true. This axiom states that for any non-empty set X and any binary relation R on X whose domain is the whole of X (i.e. for every $x \in X$ there exists some $y \in X$ with xRy) there exists a sequence with $x_n R x_{n+1}$. This generalizes the *countable axiom of choice* in the sense that it also allows to choose the elements recursively.
The principle of dependent choices seems to be more appropriate for mathematical models of physical reality than the more powerful axiom of choice, and the proofs based on it are all 'countable constructive'. Moreover, the use of this weaker axiom has the pleasant consequence that it leads to no contradiction to assume (provided that Solovay's model [44] exists) that all subsets of the real numbers are Lebesgue measurable ([44], see also [23]) and that all linear operators in Banach spaces (with full domain of definition) are bounded [17], [47]. However, we will not make use of this result (although a goal of Sect. 5.1 will be to justify the latter assumption).

We explicitly remark that the principle of dependent choices ensures that any infinite subset contains a countable subset. Thus there's no ambiguity by speaking of uncountable sets [24].

Remark 2.1.1. By using the principle of dependent choices instead of the axiom of choice most positive results of analysis remain unchanged. An important exception is Hahn-Banach's extension theorem, which we therefore will not apply. However, in many cases this theorem remains true anyway, e.g. in separable spaces [18, II.16 c)].

We recall the definition of ideal spaces. By a function x we mean in this text either a mapping from a measure space into the (extended) real line (we will just call them *real functions*) or in some nontrivial Banach space. By $|x(s)|$ we mean the absolute value of $x(s)$, resp. the norm of $x(s)$.

Definition 2.1.1. *A normed linear space X of (classes of) measurable functions with the property that the relation $|y| \leq |x|$ a.e. for some $x \in X$ and measurable y implies that $y \in X$ with $\|y\| \leq \|x\|$, is called* pre-ideal space. *A complete pre-ideal space X is called* ideal space.

In literature, ideal spaces are sometimes called *Banach function spaces* [51], and pre-ideal spaces are also known as *(normed) Köthe spaces* [48]. But be aware that sometimes also additional requirements are put on those spaces!

The following proposition implies that we may always assume that elements of pre-ideal spaces take only finite values [48, §63 Theorem 1]. Thus we may replace the extended real line by the Banach space $\mathbb{R}$.

Proposition 2.1.1. *If X is a pre-ideal space, then each $x \in X$ is finite a.e.*

Proof. Otherwise, fix an $x \in X$ where $E = \{s : |x(s)| = \infty\}$ has positive measure. Then for any $c > 0$ by $|c\chi_E| \leq |x|$

$$\|x\| \geq \|c\chi_E\| = c\,\|\chi_E\|\,,$$

a contradiction. □

Given a pre-ideal space X of real functions, and an arbitrary Banach space Y, there corresponds a pre-ideal space X_Y of Y-valued functions, defined by

$$x \in X_Y \iff x \text{ measurable and } |x| \in X,\ \|x\|_{X_Y} = \|\,|x|\,\|_X\,.$$

The converse is also true: If X_Y is a pre-ideal space of Y-valued functions, choose $e \in Y$, $|e| = 1$, and define X by

$$x \in X \iff xe \in X_Y,\ \|x\|_X = \|xe\|_{X_Y}\,.$$

Since this process is one-to-one and onto, we will call X the *real form of* X_Y. Usually, we will denote X and X_Y by the same symbol X, and will not distinguish between them. If we use symbols like $x \geq 0$ or $\sup x$, we will mean the real form of course. If we only consider the real form, we will call X *real-valued*, otherwise Y*-valued*.

We emphasize that for $Y \neq \mathbb{R}$ the above definition of ideal spaces is not as general as sometimes in literature. A more general definition (which we will only use in rare cases) is:

Definition 2.1.2. *Let Y be some Banach space over $\mathbb{K} = \mathbb{R}$ or $\mathbb{K} = \mathbb{C}$, and S be some measure space. A* pre-ideal* space *X is a normed linear space of (classes of) measurable functions $S \to Y$ with the property that for any $x \in X$ and any essentially bounded measurable function $y : S \to \mathbb{K}$ we have that $yx \in X$ and $\|yx\| \leq \|y\|_{L_\infty(S)} \|x\|$. If X is complete, it is called* ideal* space.

For pre-ideal* spaces there exists no canonical real form.

Simple examples for ideal spaces are the well-known l_p or L_p spaces or, more general, the Lebesgue-Bochner spaces $L_p(S, Y)$ for $1 \leq p \leq \infty$.
In the example of Orlicz spaces we will show the difference between pre-ideal and pre-ideal* spaces:

Example 2.1.1. Let Φ be a *Young function*, i.e. convex, even, and $\Phi(0) = 0$ with values in $[0, \infty]$ such that Φ is neither 0 nor ∞ a.e. Let S be a measure space, and Y a Banach space. Then the *Orlicz space* L_Φ consists of all measurable functions $x : S \to Y$ with finite (Luxemburg) norm

$$\|x\| = \inf\{\alpha > 0 : \int_S \Phi\left(\frac{x(s)}{\alpha}\right) ds \leq 1\}. \tag{2.1}$$

L_Φ is an ideal space, if and only if $\Phi(u)$ depends only on $|u|$, i.e. if and only if you may replace (2.1) by

$$\|x\| = \inf\{\alpha > 0 : \int_S \Phi\left(\frac{|x(s)|}{\alpha}\right) ds \leq 1\},$$

where now Φ is only defined on $\mathbb{R}$ instead on Y.

One might try to define pre-ideal spaces analogously to pre-ideal* spaces.

Definition 2.1.3. *Let Y be a Banach space, S be a measure space, and X be a normed linear space of measurable functions $S \to Y$. Denote by $\mathfrak{L}(Y)$ the space of bounded linear mappings $Y \to Y$ with the operator norm. Then we say that*

1. *X is an* $L_\infty(S, \mathbb{K})$-ideal, *if for any $x \in X$ and any measurable $y : S \to \mathbb{K}$, satisfying $|y(s)| \leq 1$ for almost all $s \in S$, we have that $z(s) = y(s)x(s)$ belongs to X with $\|z\| \leq \|x\|$.*

2. *X is an $L_\infty(S, \mathfrak{L}(Y))$-ideal, if for any $x \in X$ and any measurable $y : S \to \mathfrak{L}(Y)$, satisfying $|y(s)|_{\mathfrak{L}(Y)} \leq 1$ for almost all $s \in S$, we have that $z(s) = y(s)x(s)$ belongs to X with $\|z\| \leq \|x\|$.*

The name is motivated by the following

Remark 2.1.2. Let B be a nontrivial open or closed ball of X with center 0. Then X is an $L_\infty(S, \mathfrak{L}(Y))$-ideal ($L_\infty(S, \mathbb{K})$-ideal), if and only if B is closed w.r.t. multiplication by any function y of the unit ball U of $L_\infty(S, \mathfrak{L}(Y))$ (resp. $L_\infty(S, \mathbb{K})$).
In fact, if B is a closed ball, this is obvious. But if B is an open ball, which is closed w.r.t. multiplication by functions of U, then so is its closure $\overline{B}$, because for any $x \in \overline{B}, y \in U, 0 < \lambda < 1$ we have $\lambda x \in B$, whence $\lambda(yx) = y(\lambda x) \in B$, and thus $yx \in \overline{B}$.

The notions of Definition 2.1.3 explain the name of ideal* spaces: A space X is an $L_\infty(S, \mathbb{K})$-ideal, if and only if it is a pre-ideal* space.

However, although obviously any pre-ideal space is an $L_\infty(S, \mathfrak{L}(Y))$-ideal, it is not clear, whether any $L_\infty(S, \mathfrak{L}(Y))$-ideal in turn is a pre-ideal space. Thus we define:

Definition 2.1.4. *Let S be a measure space. We say a Banach space Y has the* characterization property, *if for any normed linear space X of measurable functions the following properties are equivalent:*

1. *X is a pre-ideal space, and*
2. *X is an $L_\infty(S, \mathfrak{L}(Y))$-ideal.*

Observe that the characterization property depends not only on the Banach space Y, but also on the underlying measure space S. But usually we will keep S fixed.

One might guess that the characterization property is equivalent to the fact that for any two measurable functions $x, z : S \to Y$ with $|z| \leq |x|$ there exists a measurable $y : S \to \mathfrak{L}(Y)$, $|y| \leq 1$ with $z(s) = y(s)x(s)$. Surprisingly, this is not true, as we shall see by a counterexample. A more careful examination shows that one has to allow an error $\varepsilon > 0$:

Proposition 2.1.2. *Let S be a measure space. A Banach space Y has the characterization property, if and only if for any measurable $x, z : S \to Y$ with $|z| \leq |x|$ and any $\varepsilon > 0$ there exists a measurable $y : S \to \mathfrak{L}(Y)$, $|y| \leq 1 + \varepsilon$ satisfying $z(s) = y(s)x(s)$.*

Proof. We first show sufficiency: Let X be an $L_\infty(S, \mathfrak{L}(Y))$-ideal. We have to show that X is a pre-ideal space. Thus, let $x \in X$ and a measurable $z : S \to Y$ satisfying $|z| \leq |x|$ be given. We have to show that $z \in X$ and $\|z\| \leq \|x\|$. By assumption, to any $\varepsilon > 0$ there exists a measurable $y_\varepsilon : S \to \mathfrak{L}(Y)$, $|y_\varepsilon| \leq 1$

and $(1+\varepsilon)^{-1}z(s) = y_\varepsilon(s)x(s)$. This implies that $z_\varepsilon = (1+\varepsilon)^{-1}z$ belongs to X with $\|z_\varepsilon\| \leq \|x\|$. Thus we also have $z \in X$, $\|z\| \leq \|x\|$.
Now we prove necessity: Let a measurable $x : S \to Y$ be given. Let X consist of all functions $z : S \to Y$, which may be written in the form $z(s) = y(s)x(s)$ with measurable $y : S \to \mathfrak{L}(Y)$. Each $z \in X$ is measurable by Theorem A.1.2. X becomes a normed linear space with norm

$$\|z\| = \inf\{\operatorname*{ess\,sup}_{s \in S} |y(s)| \mid y : S \to \mathfrak{L}(Y) \text{ is measurable with } z(s) = y(s)x(s)\}.$$

In fact, by $|z(s)| \leq |y(s)|\,|x(s)|$ we see that $\|z\| = 0$ implies $z = 0$, and to check the triangle inequality, let $z_i(s) = y_i(s)x(s)$ with $|y_i(s)| \leq \|x_i\| + \varepsilon$ for $i = 1, 2$ and observe that $\|z_1 + z_2\| \leq \operatorname{ess\,sup} |y_1(s) + y_2(s)| \leq \|z_1\| + \|z_2\| + 2\varepsilon$.
Now, if Y has the characterization property, X must be a pre-ideal space. But this means that any measurable $z : S \to Y$ with $|z| \leq |x|$ must also belong to X with $\|z\| \leq \|x\| \leq 1$. By definition of X this means that for any $\varepsilon > 0$ there exists a measurable $y : S \to \mathfrak{L}(Y)$ with $z(s) = y(s)x(s)$ and $|y| \leq 1 + \varepsilon$. □

The difficulty in the application of Proposition 2.1.2 lies of course in the fact that y has to be measurable: If $\mathfrak{L}(Y)$ is not separable, measurability is quite restrictive.
If we set $\varepsilon = 0$ in Proposition 2.1.2 we get the condition we mentioned before, which is sufficient for the characterization property:

Definition 2.1.5. *Let S be a measure space. A Banach space Y has the* strong characterization property, *if for any measurable $x, z : S \to Y$ with $|z| \leq |x|$ there exists a measurable $y : S \to \mathfrak{L}(Y)$, $|y| \leq 1$ satisfying $z(s) = y(s)x(s)$.*

It is not obvious, whether the characterization property is really weaker than the strong characterization property, since, given $|z| \leq |x|$ the characterization property implies that there exists a sequence y_n with $z(s) = y_n(s)x(s)$ and $|y_n| \leq 1 + n^{-1}$. But it is not clear, whether a *convergent* sequence with these properties can be found (and indeed, later we will give a counterexample).

The strong characterization property is simpler to study, since it turns out that it is connected with properties of the duality map:

Definition 2.1.6. *Let Y be a Banach space, and Y^* its dual space. The* duality map *F is a (multivalued) mapping of Y into Y^*, defined by $y^* \in F(y)$, if $y^*(y) = |y|^2$ and $|y^*| = |y|$.*

We call F *nontrivial*, if $F(y) \neq \emptyset$ for at least one $y \neq 0$, and *fully defined*, if $F(y) \neq \emptyset$ for any $y \in Y$.

Remark 2.1.3. Recall that Hahn-Banach's theorem need not hold for Y (Remark 2.1.1). However, if it does hold for Y (e.g. if Y is separable), then F is fully defined [1, 4.4.1]. Also in Hilbert spaces F is fully defined (Here we even have $F = I$).
We remark however that (without the axiom of choice!) it may happen in $Y = L_\infty([0,1])$ that the duality map of $y(s) = s$ is empty: It is easy to see that it contains no integral functional (see Definition 3.4.2), and in Solovay's model *all* bounded functionals on Y are integral functionals (Theorem 3.5.6).

The following theorem allows us to reduce the problem of finding a measurable function with values in $\mathfrak{L}(Y)$ to finding a measurable function with values in Y^*, which usually is much easier of course:

Theorem 2.1.1. *Let S be a measure space, and Y be a Banach space with a nontrivial duality map F. Then the strong characterization property is equivalent to the fact that for any measurable $x : S \to Y$ there exists a measurable selection of $F \circ x$.*

Proof. We first prove sufficiency. Let $x, z : S \to Y$ be measurable, $|z| \leq |x|$. Let G be a measurable selection of $F \circ x$, and define $y : S \to \mathfrak{L}(Y)$ by

$$y(s) = \begin{cases} z(s)\,|x(s)|^{-2}\,G(s) & \text{if } x(s) \neq 0, \\ 0 & \text{if } x(s) = 0. \end{cases} \tag{2.2}$$

Then y is measurable, $|y(s)| \leq 1$, and $y(s)x(s) = z(s)$.
Now we prove necessity. Assume, Y has the strong characterization property. Let $x : S \to Y$ be measurable. Since F is nontrivial, there exist $e \in Y$, $e^* \in Y^*$, satisfying $|e| = |e^*| = 1$ and $e^*(e) = 1$. We apply the strong characterization property for $z(s) = |x(s)|\,e$ and find that there exists a measurable $y : S \to \mathfrak{L}(Y)$, $|y| \leq 1$ with $|x(s)|\,e = y(s)x(s)$.
Now define $G : S \to Y^*$ by $G(s) = |x(s)|\,e^* y(s)$. Then G has the properties $G(s)x(s) = |x(s)|^2$ and $|G(s)| \leq |x(s)|$. But this means that $G(s) \in F(x(s))$. Thus G is a measurable selection of $F \circ x$. □

We make some remarks on Theorem 2.1.1.
Call a function f in Banach spaces *sup-measurable*, if $f \circ x$ is measurable for any measurable x.
Since the duality map is a map, if and only if the dual space is strictly convex [12, Proposition 12.3(b)], Theorem 2.1.1 immediately implies:

Corollary 2.1.1. *Let a Banach space Y have a strictly convex dual space. Assume, the (single-valued) duality map F is nontrivial. Then Y has the strong characterization property, if and only if F is sup-measurable.*

Theorem 2.1.1 implies that a Banach space has the strong characterization property, if its duality map has a sup-measurable selection.

Observe that continuous functions are sup-measurable. The duality map is continuous, if the dual space is uniformly convex [12, Prop. 12.3(c)]. Thus we have the

Corollary 2.1.2. *Let a Banach space Y have a uniformly convex dual space and a fully defined duality map. Then Y has the strong characterization property.*

In particular, any Hilbert space has the strong characterization property. But there exist also examples of spaces with the strong characterization property, which have no strictly convex dual space: $Y = \mathbb{R}^n$ with the maximum norm or with the sum norm has a sup-measurable (although not continuous) selection of the duality map, which is easily checked (observe, that sgn is sup-measurable).

Finally, we give a natural example of a separable space without the strong characterization property:

Example 2.1.2. Let $S = [0,1]$, $Y = L_1(S)$. We show that Y does not have the strong characterization property by using Theorem 2.1.1. Define $x : S \to Y$ by

$$x(t)(s) = \begin{cases} 1 & \text{if } s \leq t, \\ -1 & \text{if } s > t. \end{cases}$$

x is measurable by Theorem 4.4.2. Let F be the duality map, and $y^*(t) \in F(x(t))$. By the following Lemma 2.1.1, we must have

$$y^*(t)y = \int_0^t y(s)ds - \int_t^1 y(s)ds.$$

For $t \leq \tau$ we have $[y^*(\tau) - y^*(t)]\, y = 2\int_t^\tau y(s)ds$, which implies $|y^*(t) - y^*(\tau)| = 2$ for $t \neq \tau$. Thus by Lemma A.1.1, y^* is not essentially separable valued, whence not measurable.

Lemma 2.1.1. *Let S be a σ-finite measure space, and $Y = L_1(S, \mathbb{K})$. The value $F(y)$ of the duality map consists of all $y^* \in Y^*$ of the form*

$$y^*(x) = \|y\|_Y \int_S z(s)x(s)ds, \tag{2.3}$$

where $z : S \to \mathbb{K}$ is measurable, satisfying $|z| \leq 1$ and $z(s)y(s) = |y(s)|$.

Proof. We show that any $y^* \in F(y)$ must have this form. By $y^* \in Y^*$, it is clear that (2.3) holds for some measurable $z : S \to \mathbb{K}$ with $\|y\|_Y \operatorname{ess\,sup} |z(s)| = \|y^*\|_{Y^*}$. But since $y^* \in F(y)$, this implies (for $y \neq 0$) that

$$\int_S z(s)y(s)ds = \|y\|_Y^{-1}\, y^*(y) = \|y\|_Y = \int_S |y(s)|\, ds,$$

and $\operatorname{ess\,sup} |z(s)| \le 1$. From this the statement is obvious (cf. e.g. [40, Theorem 1.39(c)]). □

If we want to state an analogous result to Theorem 2.1.1 for the characterization property, we have to widen the range of the duality map:

Definition 2.1.7. *Let Y be a Banach space, Y^* its dual space, and $\varepsilon > 0$. The* ε-duality map *is a multivalued function from Y into Y^*, defined by $y^* \in F(y)$, if $y^*(y) = |y|^2$ and $|y^*| \le |y|(1+\varepsilon)$.*

Theorem 2.1.2. *Let S be a measure space, and Y be a Banach space with nontrivial ε-duality maps F_ε $(\varepsilon > 0)$. Then Y has the characterization property, if and only if for any $\varepsilon > 0$ and any measurable $x : S \to Y$ there exists a measurable selection of $F_\varepsilon \circ x$.*

Proof. For both directions we use Proposition 2.1.2 and continue similarly as in the proof of Theorem 2.1.1.
For sufficiency, let $x, z : S \to Y$ be measurable, $|z| \le |x|$ and $\varepsilon > 0$ be given. Let G be a measurable selection of $F_\varepsilon \circ x$, and define y by (2.2). Then y is measurable, $|y(s)| \le 1 + \varepsilon$ and $y(s)x(s) = z(s)$.
For necessity assume, Y has the strong characterization property. Let a measurable $x : S \to Y$ and $\varepsilon > 0$ be given. Let $0 < \delta < \varepsilon$. Since F_δ is nontrivial, there exist $e \in Y$, $e^* \in Y^*$, satisfying $|e| = 1$, $|e^*| \le 1 + \delta$ and $e^*(e) = 1$. Applying Proposition 2.1.2 for $z(s) = |x(s)|\, e$, we find that there exists a measurable $y : S \to \mathfrak{L}(Y)$, $|y| \le (1+\delta)^{-1}(1+\varepsilon)$ with $|x(s)|\, e = y(s)x(s)$. Now define $G : S \to Y^*$ by $G(s) = |x(s)|\, e^* y(s)$. Then G has the properties $G(s)x(s) = |x(s)|^2$ and $|G(s)| \le |x(s)|(1+\varepsilon)$. But this means that $G(s) \in F_\varepsilon(x(s))$. Thus G is a measurable selection of $F_\varepsilon \circ x$. □

We remark that for the necessary part we even showed a bit more: If there exist $0 < \delta < \varepsilon$ and a measurable x, such that F_δ is nontrivial and $F_\varepsilon \circ x$ contains no measurable selection, then Y may not have the characterization property. This sharpening is of course only of interest, if you do not assume the axiom of choice (as we do), because otherwise the condition that each F_δ is nontrivial is always satisfied.

However, even with our axioms the assumption that F_ε is nontrivial for any $\varepsilon > 0$ is not restrictive (for ideal spaces see e.g. Corollary 3.4.8).
It is satisfied, for example, if the canonical embedding of Y into Y^{**} is norm preserving, i.e. if Y has the following property:

Definition 2.1.8. *A Banach space Y has the* bidual property, *if*

$$\|y\|_Y = \sup_{\|y^*\|_{Y^*} \le 1} |y^*(y)| \qquad (y \in Y),$$

where Y^* denotes the dual space of Y.

The bidual property is equivalent to the fact that for any $\varepsilon > 0$ the ε-duality map is fully defined.
Y has the bidual property and the mentioned supremum is even a maximum, if and only if the duality map of Y is fully defined. Thus recall Remark 2.1.3. In particular, any separable Banach space and any Hilbert space has the bidual property.
Also the most important nonseparable spaces used in practice have the bidual property. For example, for almost perfect ideal spaces over σ-finite measure spaces like $L_\infty([0,1])$ one might use Corollary 3.4.8.
Observe that this is in contrast to Remark 2.1.3: For $Y = L_\infty([0,1])$ the duality map of $y(s) = s$ may be empty (without the axiom of choice!). Thus the bidual property is less restrictive than the assumption that the duality map is fully defined. However, (without the axiom of choice) it can not be disproved that there exist spaces without the bidual property, as we will see in Corollary 3.5.4.

Now we will state our main result about the characterization property, namely that it holds in almost any case. The proof is based on the technical Theorem A.2.1.

Theorem 2.1.3. *Let S be a σ-finite measure space, and Y have the bidual property. Then Y has the characterization property.*

Proof. We use Theorem 2.1.2. Let $\varepsilon > 0$ and a measurable $x : S \to Y$ be given. We have to prove that $F_\varepsilon \circ x$ contains a measurable selection G, where F_ε denotes the ε-duality map of Y.
Let $\alpha = \varepsilon/(1+\varepsilon)$. Corollary A.2.2, applied for $z(s) = \alpha\,|x(s)|$ implies that there exists a measurable $y : S \to Y^*$ with $|y| = 1$ and

$$y(s)x(s) \geq |x(s)| - \alpha\,|x(s)| = (1+\varepsilon)^{-1}\,|x(s)|\,.$$

Hence the function

$$G(s) = \begin{cases} \frac{|x(s)|^2}{y(s)x(s)}y(s) & \text{if } x(s) \neq 0, \\ 0 & \text{if } x(s) = 0, \end{cases}$$

is defined and measurable and satisfies $|G(s)| \leq |x(s)|\,(1+\varepsilon)$ and $G(s)x(s) = |x(s)|^2$. Thus G is a measurable selection of $F_\varepsilon \circ x$. $\square$

We remark that if we assume the axiom of choice the second assumption of Theorem 2.1.3 is always satisfied, and the first can be essentially weakened (for example it suffices that S has the direct sum property, see Remark A.2.1). Thus in this case *any* Banach space has the characterization property for a very large class of measure spaces (maybe even for all).

Anyway, Theorem 2.1.3 covers the cases important in practice. It also shows (see e.g. Corollary 3.4.8) that for $S = [0,1]$ the space $Y = L_1(S)$ has the characterization property (although it has not the strong characterization property, as we have seen in Example 2.1.2). Hence this space is the counterexample we promised in the beginning.

2.2 Extended Convergence and the Support of Functions

We will often need the fact that $x_n \to x$ in some pre-ideal space and $x_n \to y$ in some other sense implies $x = y$. This is based on the following

Lemma 2.2.1. *Let X be a pre-ideal space, $x_n \to x$ in X. Fix some representation of x_n and x. Then for almost all s there exists a subsequence, such that $x_{n_k}(s) \to x(s)$.*

Proof. If for some s there is no such subsequence, then there exist natural numbers n, m, such that s lies in

$$D_{nm} = \bigcap_{k=n}^{\infty} \{s : |x(s) - x_k(s)| \geq m^{-1}\}. \tag{2.4}$$

Since $y = m^{-1}\chi_{D_{nm}}$ satisfies $|y| \leq |x - x_k|$ for all $k \geq n$, hence $y \in X$ and $\|y\| \leq \|x - x_k\| \to 0$, we have $\mathrm{mes} D_{nm} = 0$. Thus the union of all D_{nm} has measure zero. □

From Lemma 2.2.1 we may conclude of course that $x_n \to x$ in X and $x_n \to y$ a.e. imply that $x = y$ a.e. (Theorem 2.2.3). But in many of our applications (e.g. in the proof that the Riesz-Fischer property is necessary for completeness, see Theorem 3.2.1) we do not know a priori that $|x_n(s)|$ is bounded for almost all s.

Thus we introduce a notion of convergence, which also covers the case $|x_n(s)| \to \infty$. Since we assume that x_n takes values in a Banach space Y, it is not sufficient just to consider the extended real line. We will say that a sequence of measurable functions x_n *converges (a.e.) in the extended sense on S* to some function x, if for all (almost all) $s \in S$ we either have $x(s) = \lim x_n(s)$, or $|x_n(s)| \to \infty$, in which case we write $x(s) = \infty$. The statement that x belongs to some pre-ideal* space then means by definition also that x is finite almost everywhere.
To be able to speak also of convergence in measure in the extended sense, we have to consider an appropriate distance on $Y \cup \{\infty\}$. For example, you may choose the chordal metric (which was suggested by P. P. Zabrejko 1995), defined by

$$\begin{aligned}
\hat{d}(x,y) &:= \frac{\sqrt{|x(1+|y|^2) - y(1+|x|^2)|^2 + (|x|^2 - |y|^2)^2}}{(1+|x|^2)(1+|y|^2)} \qquad (x, y \in Y), \\
\hat{d}(x,\infty) &:= \hat{d}(\infty, x) := \frac{1}{\sqrt{1+|x|^2}} \quad (x \in X), \qquad \hat{d}(\infty,\infty) := 0.
\end{aligned} \tag{2.5}$$

We remark that in the Hilbert space case a straightforward calculation shows that the term under the squareroot may be simplified to $(1 + |x|^2)(1 + |y|^2)|x - y|^2$. Thus in this case we have

$$\hat{d}(x,y) = \frac{|x-y|}{\sqrt{1+|x|^2}\sqrt{1+|y|^2}} \qquad (x,y \in Y)$$

(i.e. especially on $Y = \mathbb{R}^2 \cong \mathbb{C}$ this is the usual chordal metric).

Proposition 2.2.1. (2.5) *defines a metric on $Y \cup \{\infty\}$. If restricted to Y, this metric is equivalent to the metric induced by the norm. Moreover, in this metric a sequence $y_n \in Y$ converges to ∞, if and only if $|y_n| \to \infty$.*

Proof. Let $Z = Y \times \mathbb{R}$ be the real vector space with the natural operations (if Y is complex, we just restrict the multiplication to real scalars). On Z we may define a norm by

$$\|(x,r)\| := \|(|x|,|r|)\|_2 = \sqrt{|x|^2+r^2} \qquad (x \in Y,\ r \in \mathbb{R}).$$

To see that this satisfies the triangle inequality, observe that $\|(x+y,r+s)\| \le \|(|x|+|y|,|r|+|s|)\|_2$ and use the triangle inequality for the Euclidean norm. Now let $S = \{(x,r) \in Z : \|(x,r)-(0,1/2)\| = 1/2\}$ be the sphere in Z with 'south pole' $(0,0)$ and 'north pole' $N = (0,1)$, and $\dot{S} = S \setminus \{N\}$. Define $\varphi : Y \to Z$ and $\psi : \dot{S} \to Y$ by

$$\varphi(x) = \frac{1}{1+|x|^2}(x,|x|^2), \quad \psi(x,r) = \frac{x}{1-r}.$$

Since $\|\varphi(x)-(0,1/2)\|^2 = (1+|x|^2)^{-2}(|x|^2+(|x|^2/2-1/2)^2) = 1/4$, we have $\varphi : Y \to S$, even $\varphi : Y \to \dot{S}$. Moreover,

$$(\psi \circ \varphi)(x) = \frac{x}{1+|x|^2-|x|^2} = x \qquad (x \in Y),$$

and for $(x,r) \in \dot{S}$ we have by $r \neq 1$ and $|x|^2 + (r-1/2)^2 = 1/4$ that

$$\begin{aligned}(\varphi \circ \psi)(x,r) &= \frac{1}{(1-r)^2+|x|^2}(x(1-r),|x|^2) \\ &= \frac{1}{[|x|^2+(r-1/2)^2]-r+3/4}(x(1-r),\tfrac{1}{4}-(r-\tfrac{1}{2})^2) = (x,r)\end{aligned}$$

Thus φ is one-to-one and onto $\dot{S}$ with inverse mapping ψ. Both functions φ and ψ are continuous, as can be seen by

$$\varphi(x)-\varphi(y) = \frac{1}{(1+|x|^2)(1+|y|^2)}((x-y)(1+|y|^2)-(|x|^2-|y|^2)y,|x|^2-|y|^2)$$

and

$$\psi(x,r)-\psi(y,s) = \frac{x-y}{1-r} + \frac{(r-s)y}{(1-r)(1-s)} \qquad ((x,r),(y,s) \in \dot{S}).$$

Thus φ is a homeomorphism of Y and $\dot{S}$. In particular,

$$\hat{d}(x,y) = \|\varphi(x) - \varphi(y)\|$$

defines a metric on Y, which is equivalent to the given metric induced by the norm. Moreover, allowing $\varphi(\infty) := N = (0,1)$, we must have an extension of this metric on $Y \cup \{\infty\}$, since $\varphi : Y \cup \{\infty\} \to S$ still is one-to-one. A straightforward calculation shows that this is the metric given by (2.5). The last statement now follows immediately by (2.5). □

Using Proposition 2.2.1 we have that $x_n \to x$ (a.e.) in the extended sense, if and only if $\hat{d}(x(s), x_n(s)) \to 0$ (a.e.). We say that $x_n \to x$ *uniformly on* S *in the extended sense*, if

$$\lim_{n\to\infty} \sup_{s\in S} \hat{d}(x(s), x_n(s)) \to 0.$$

Analogously, we define that a sequence of measurable functions x_n *converges in measure in the extended sense on* S to some measurable function x, if for all $c > 0$ we have

$$\lim_{n\to\infty} \text{mes}\{s \in S : \hat{d}(x(s), x_n(s)) \geq c\} = 0.$$

Riesz's and Egorov's theorems carry over:

Theorem 2.2.1. *Let $x_n \to x$ in measure in the extended sense. Then there exists a subsequence such that $x_{n_k} \to x$ a.e. in the extended sense.*

Theorem 2.2.2. *Let x_n be measurable, $x_n \to x$ a.e. in the extended sense on some set E of finite measure. Then for any $\varepsilon > 0$ there exists some measurable $D \subseteq E$ with* $\text{mes}(E \setminus D) \leq \varepsilon$, *such that $x_n \to x$ uniformly on D in the extended sense. Furthermore, then $x_n \to x$ in measure in the extended sense on E.*

For the proofs just apply the classical theorems to the measurable and everywhere finite sequence $y_n(s) = \hat{d}(x(s), x_n(s))$ with limit function 0.

Definition 2.2.1. *The* support *of a function x, $\text{supp}x$, is the set of all s, for which $x(s) \neq 0$ (defined up to null sets).*

Definition 2.2.2. *A subset E of some measure space has the* finite subset property, *if any subset of E with positive measure contains a subset of positive but finite measure.*

Any set with σ-finite measure has the finite subset property, but the converse is not true: The counting measure on some uncountable set E has the finite subset property, but is not σ-finite.

Theorem 2.2.2 implies a result, which looks similar to Lemma 2.2.1:

Lemma 2.2.2. *Let X be a pre-ideal space. Let x_n be a fixed representation of a sequence in X, which is bounded in norm and for which each* suppx_n *has the finite subset property. Then for almost all s there exists a subsequence x_{n_k}, such that $x_{n_k}(s)$ is bounded.*

Proof. The set $E = \{s : |x_n(s)| \to \infty\}$ lies in the Lebesgue-extension of the measure space, since

$$E = \bigcap_{k=1}^{\infty} \bigcup_{N=1}^{\infty} \bigcap_{n=N}^{\infty} \{s : |x_n(s)| \geq k\}.$$

We have to prove that mes$E = 0$. Otherwise, since $E \subseteq \bigcup \text{supp} x_n$ has the finite subset property, E contains a measurable subset $F \subseteq E$ of positive but finite measure. By Theorem 2.2.2, F again contains a set $D \subseteq E$ of positive measure, such that $|x_n| \to \infty$ uniformly on D. In particular, for any k there exists n_k with $|x_{n_k}| \geq |k\chi_D|$, i.e. $\|x_{n_k}\| \geq k\,\|\chi_D\|$ in contradiction to the boundedness of $\|x_{n_k}\|$. □

The classical dominated convergence theorem of Lebesgue states for $L_1(S)$ that a dominated sequence converges in norm, if it converges a.e. But instead of convergence a.e. you may also consider convergence in measure. Or even convergence in measure on each set of finite measure. To cover all these cases without always repeating them, we define:

Definition 2.2.3. *A sequence x_n of measurable functions* converges extended *to some x, if one of the following conditions is true:*

1. *$x_n \to x$ a.e. in the extended sense, or*
2. *$x_n \to x$ in measure in the extended sense, or*
3. *$x_n \to x$ in measure in the extended sense on every set of finite measure, and* supp$x$ *and each* suppx_n *is σ-finite.*

We say that x_n converges extended* *to x, if it converges extended, or if*

3.* *$x_n \to x$ in measure in the extended sense on every set of finite measure, and* supp$x$ *and each* suppx_n *has the finite subset property.*

The limit function is measurable and unique:

Proposition 2.2.2. *Let $x_n \to x$ extended* and $x_n \to y$ extended*. Then x is measurable, and $x = y$ a.e.*

Proof. If x is not measurable, there exists a set E of finite measure, such that x is not the limit of simple functions on E, i.e. not measurable on E. But this is not possible, because $x_n \to x$ in measure on E in the extended sense. For the uniqueness of the limit, it suffices to consider the case that supp$x$ and all suppx_n have the finite subset property. But then either also supp$y$ has the finite subset property, or $x_n \to y$ a.e. or in measure in the extended sense.

But in the second case we have $\operatorname{supp} y \subseteq \bigcup \operatorname{supp} x_n$, hence again $\operatorname{supp} y$ has the finite subset property. In both cases the set M of all s with $x(s) \neq y(s)$ has by $M \subseteq \operatorname{supp} x \cup \operatorname{supp} y$ the finite subset property. Thus, if M has positive measure, it contains a subset E of positive but finite measure, which is not possible, because $x_n \to x$ in measure on E in the extended sense. □

For later reference we note the following observation:

Corollary 2.2.1. *Let x_n be real and measurable. If $x_n \to x$ extended* and if the sequence $x_n(s)$ is monotone for almost all s, then $x_n \to x$ a.e. in the extended sense.*

Proof. For all s, for which $x_n(s)$ is monotonically increasing, define $y(s) = \sup x_n(s)$, for all other, let $y(s) = \inf x_n(s)$. Then $x_n \to y$ a.e. in the extended sense. But since now Proposition 2.2.2 implies $x = y$, this is the statement. □

By Theorem 2.2.1 we have

Lemma 2.2.3. *If $x_n \to x$ extended, then there exists a subsequence with $x_{n_k} \to x$ a.e. in the extended sense.*

Proof. By Theorem 2.2.1 it suffices to consider the last case. Let $\bigcup \operatorname{supp} x_n \cup \operatorname{supp} x = \bigcup S_k$ with $\operatorname{mes} S_k < \infty$. By Theorem 2.2.1 for some subsequence we have $x_{n_{k,1}} \to x$ a.e. in the extended sense on S_1, and for a subsequence of this $x_{n_{k,2}} \to x$ a.e. in the extended sense on S_2, and so on. The diagonal sequence $x_{n_{k,k}}$ converges on every S_k a.e. in the extended sense to x. □

For extended* convergence we have only analogously Lemma 2.2.1:

Lemma 2.2.4. *If $x_n \to x$ extended*, then for almost all s there exists a subsequence with $x_{n_k}(s) \to x(s)$.*

Proof. By Lemma 2.2.3 it suffices to consider the last case. Let M be the set of all s, for which such a subsequence does not exist. M lies in the Lebesgue-extension of the measure space, since $M = \bigcup_{n,m} D_{nm}$, where D_{nm} is given by (2.4). Assume, M has positive measure. Since $M \subseteq \bigcup \operatorname{supp} x_n \cup \operatorname{supp} x$ has the finite subset property, there exists some $E \subseteq M$ of positive but finite measure. But this is not possible by Theorem 2.2.1. □

Lemma 2.2.5. *Let X be a pre-ideal* space, $x_n \to x$. Then* $\operatorname{supp} x \subseteq \bigcup \operatorname{supp} x_n$.

Proof. Let $M = \bigcup \operatorname{supp} x_n$, $P_M x(s) = \chi_M(s)x(s)$. Then

$$\|x_n - P_M x\| = \|P_M(x_n - x)\| \le \|x_n - x\| \to 0$$

implies $x_n \to P_M x$, i.e. $x = P_M x$. □

Corollary 2.2.2. *Let X be an ideal* space over S, $E \subseteq S$ be measurable. Then the restriction X_E of X to functions, which vanish outside E, is an ideal* space.*

Proof. X_E is closed by Lemma 2.2.5. □

Now we are able to state and prove our first goal:

Theorem 2.2.3. *Let X be a pre-ideal space, $x_n \to x$ in X, and $x_n \to y$ extended*. Then $x = y$ a.e.*

Proof. Assume, $E = \{s : x(s) \neq y(s)\}$ has positive measure. We claim that there exists a subset $D \subseteq E$ of positive measure and a subsequence x_{n_k}, which converges a.e. on D to y in the extended sense. Then Lemma 2.2.1 yields the contradiction $x = y$ a.e. on D.
If $x_n \to y$ extended, we may choose $D = E$ and use Lemma 2.2.3. Otherwise, we may assume that $\operatorname{supp} y$ and all $\operatorname{supp} x_n$ have the finite subset property. By Lemma 2.2.5 also $\operatorname{supp} x \subseteq \bigcup \operatorname{supp} x_n$ has the finite subset property. But then $E \subseteq \operatorname{supp} x \cup \operatorname{supp} y$ contains a set D of positive but finite measure. Now we have that $x_n \to y$ on D extended, and may again use Lemma 2.2.3. □

We remark that the proof of Theorem 2.2.3 substantially simplifies, if we assume that the underlying measure space has the finite subset property. Then you might either just apply the results of Sect. 3.1, or you might argue straightforwardly (for simplicity we also assume that y is finite, although it is an easy exercise to drop this now by applying Lemma 2.2.2):
Indeed, if we then have $x \neq y$ in the theorem, there exist N and a set E of positive but finite measure with $|x - y| > N^{-1}$ on E. Since $x_n \to y$ in measure on E, Egorov's theorem implies that $x_n \to y$ uniformly on a set $D \subseteq E$ of positive measure. Hence, for n large enough, we have $|y - x_n| \le |x - x_n|$ on D. Thus, if we redefine $y(s) := x(s)$ outside D, we have $\|y - x_n\| \le \|x - x_n\| \to 0$, hence $x_n \to y \neq x$ in X, a contradiction.

We emphasize that in the following corollary the finiteness of x is not assumed a priori:

Corollary 2.2.3. *Let X be an ideal space, and x_n be a Cauchy sequence in X, and $x_n \to x$ extended*. Then x is finite a.e., $x \in X$, and $\|x - x_n\| \to 0$.*

Proof. Since X is complete, we have $x_n \to y$ in X for some $y \in X$. Theorem 2.2.3 implies $x = y$. □

We remark that the previous results all fail, if in the definition of extended* convergence we drop in the last case the assumption that $\operatorname{supp} x$ and each $\operatorname{supp} x_n$ has the finite subset property, even if each point is contained in some set of finite measure:

Example 2.2.1. Let S be some uncountable set, each $E \subseteq S$ be measurable, $\operatorname{mes} E = 0$ for countable E, $\operatorname{mes} E = \infty$ otherwise. Lemma 2.2.4 then fails for $x_n = 0$, $x = 1$ and for $x_n = 1$, $x = 0$. Theorem 2.2.3 and Corollary 2.2.3 fail, if you choose $X = \{0\}$, and $y = 1$ as the limit function, or if you choose $X = L_\infty(S)$, $x_n = 1$, and $y = 0$.

Lemma 2.2.6. *Let X_Y be some Y-valued pre-ideal space, X be its real form, and $e \in Y$. Assume, for a sequence $x_n \in X$ we have that $x_n e$ converges extended* or in norm to some $y \in X_Y$. Then there exists some $x \in X$ with $y = ex$.*

Proof. By Lemma 2.2.4 resp. Lemma 2.2.1 we have that for almost all s there exists a subsequence with $x_{n_k}(s)e \to y(s)$. For those s there exists some number $x(s)$ with $y(s) = x(s)e$. Since y is measurable, also x is measurable. Thus $y \in X_Y$ implies $x \in X$. □

Lemma 2.2.7. *Let X_Y be an ideal space of Y-valued functions. Then its real form X is an ideal space.*

Proof. We prove completeness. Let $e \in Y$, $|e| = 1$. Given a Cauchy sequence $x_n \in X$, then $x_n e$ is a Cauchy sequence in X_Y, whence $x_n e \to y$ in X_Y for some $y \in X_Y$. By Lemma 2.2.6 we have $y = xe$ for some $x \in X$. Furthermore, $\|x - x_n\|_X = \|y - x_n e\|_{X_Y} \to 0$. □

We remark that the converse of Lemma 2.2.7 is also true, as we shall see soon (Theorem 3.2.1).

Corollary 2.2.4. *Let X be an ideal space, $x_n \in X$. Then there exists $u \in X$ with $\operatorname{supp} x_n \subseteq \operatorname{supp} u$ for all n.*

Proof. By Lemma 2.2.7 we may assume without loss of generality that X is real-valued. There exist $c_n > 0$ with $\sum c_n \|x_n\| < \infty$. Consider

$$u(s) = \sum_{n=1}^{\infty} c_n |x_n(s)|.$$

Since the partial sums are Cauchy in X, we have $u \in X$ by Corollary 2.2.3. □

Corollary 2.2.5. *If S is a σ-finite measure space, then $L_p(S)$ contains a nonvanishing function.*

Proof. Let $S = \bigcup S_n$ with $\text{mes} S_n < \infty$. Since $x_n = \chi_{S_n} \in L_p$, the statement follows by Corollary 2.2.4. □

As a consequence of this, one can show that any σ-finite measure can be normalized (see e.g. [50]):

Corollary 2.2.6. *Let (S, Σ, μ) be a σ-finite measure space. Then there exists a measure ν, such that (S, Σ, ν) has the same null sets, but $\nu(S) = 1$.*

Proof. By Corollary 2.2.5 there exists a strictly positive function u, which is integrable with respect to μ. Then define

$$\nu(E) = c \int_E u(s) d\mu(s) \qquad (E \in \Sigma),$$

where $c = 1/\int_S u(s) d\mu(s)$ is a norming constant. □

Definition 2.2.4. *Let M be a set of measurable functions. A measurable set E is called* support of M, $\text{supp} M$, *if it has the following two properties:*

(a) Each $x \in M$ vanishes a.e. outside E, and
(b) for each $D \subseteq E$ with $\text{mes} D > 0$ *there exists a function $x \in M$ with* $\text{mes}(D \cap \text{supp} x) > 0$.

If $\text{supp} M$ is defined, then it is determined uniquely up to sets of measure zero: In fact, assume the contrary, i.e. there exists another set E with the properties of $\text{supp} M$, such that (without loss of generality) $D = \text{supp} M \setminus E$ has positive measure. Then there exists some $x \in M$, such that $D \cap \text{supp} x$ has positive measure, whence x does not vanish a.e. outside E, a contradiction.

Observe that $\text{supp} M$ need not be defined on arbitrary measure spaces, even if M is an ideal space:

Example 2.2.2. Define a measure space S in the following way: Let $S = T \cup R \cup N$ be the union of three pairwise disjoint uncountable sets T, R, N. Call $E \subseteq S$ measurable, if not both of the sets E and $S \setminus E$ are uncountable. If E is measurable, let $\text{mes} E = \infty$ if E is uncountable, otherwise let $\text{mes} E$ be the number of elements in $E \setminus N$. Let M be the ideal space of measurable functions $x : S \to \mathbb{R}$ with $x|_R = 0$ and $\|x\| = \text{ess sup}\, |x(s)| < \infty$. Then a set $E \subseteq S$ is the support of some $x \in M$, if and only if E is at most countable and $E \subseteq T \cup N$. Hence the sets with the properties of $\text{supp} M$ are precisely all sets E with $T \subseteq E \subseteq T \cup N$. But none of these sets is measurable.

The example shows additionally that we run into serious uniqueness problems, if we drop the condition that $\mathrm{supp}M$ is measurable.

The problem in the definition of $\mathrm{supp}M$ is due to the fact that for any $x \in M$ the set $\mathrm{supp}x$ is only defined up to null sets, i.e. there are usually many sets E representing $\mathrm{supp}x$. However, in a straightforward way we might define:

Definition 2.2.5. *Let M be a set of measurable functions. If* $\mathrm{supp}M$ *exists, let* $\overline{\mathrm{supp}}M = \underline{\mathrm{supp}}M = \mathrm{supp}M$. *Otherwise let*

$$\overline{\mathrm{supp}}M = \bigcup_{x \in M} \bigcup \{E : E \textit{ is a representation of } \mathrm{supp}x\} \tag{2.6}$$

$$\underline{\mathrm{supp}}M = \bigcup_{x \in M} \bigcap \{E : E \textit{ is a representation of } \mathrm{supp}x\} \tag{2.7}$$

Obviously, $\underline{\mathrm{supp}}M \subseteq \overline{\mathrm{supp}}M$. If $\mathrm{supp}M$ does not exist, these sets are a good replacement for this, as we will see.
The previous Example 2.2.2 shows that even on complete measure spaces it may happen that none of the sets $\overline{\mathrm{supp}}M$, $\underline{\mathrm{supp}}M$, or $\overline{\mathrm{supp}}M \setminus \underline{\mathrm{supp}}M$ is measurable.
Observe that if $\mathrm{supp}M$ exists, it is no good idea to use the definitions (2.6)/(2.7): For example, if all sets consisting of a single point are null sets (e.g. for the Lebesgue measure in $\mathbb{R}$), (2.6) is always the whole measure space, while (2.7) is empty.
$\overline{\mathrm{supp}}M$ is (in the best way we may define) the 'smallest' set satisfying property (a) of Definition 2.2.4, and $\underline{\mathrm{supp}}M$ is the 'largest' set with property (b):

Lemma 2.2.8. *Let M be a set of measurable functions. Then each $x \in M$ vanishes a.e. outside* $\overline{\mathrm{supp}}M$. *Furthermore, for any measurable* $E \subseteq \underline{\mathrm{supp}}M$ *with* $\mathrm{mes}E > 0$ *there exists some $x \in M$ with* $\mathrm{mes}(E \cap \mathrm{supp}x) > 0$.

Proof. If $\mathrm{supp}M$ exists, the statement is trivial. Otherwise the first statement is also trivial. For the second statement, let $E \subseteq \underline{\mathrm{supp}}M$ be of positive measure. By (2.7) there exists some $x \in M$ such that $E \cap D \neq \emptyset$ for any representation D of $\mathrm{supp}x$. Hence $E \cap \mathrm{supp}x$ may not have measure zero, since otherwise there exists some representation D of $\mathrm{supp}x$ with $E \cap D = \emptyset$.
□

On σ-finite measure spaces, $\mathrm{supp}M$ is always defined (cf. [48, §67 Theorem 2]):

Theorem 2.2.4. *If S is a σ-finite measure space, and M is a set of measurable functions over S, then* $\mathrm{supp}M$ *exists.*

Proof. Let $S = \bigcup S_n$, mes$S_n < \infty$. Let Γ be the system of all measurable subsets $E \subseteq S$, such that each $x \in M$ vanishes a.e. on E. There exists a sequence $D_k \in \Gamma$, such that

$$\text{mes}(D_k \cap S_1) \to \sup\{\text{mes}(E \cap S_1) : E \in \Gamma\} = m.$$

Hence $E_1 = \bigcup D_k \cap S_1$ satisfies mes$E_1 \geq m$. For any $D \subseteq S_1 \setminus E_1$ of positive measure we have mes$((E_1 \cup D) \cap S_1) >$ mes$E_1 \geq m$, whence $E_1 \cup D \notin \Gamma$ by the definition of m, which implies by $E_1 \in \Gamma$ that $D \notin \Gamma$. Analogously, we can find subsets $E_n \subseteq S_n$ with $E_n \in \Gamma$, such that for any $D \subseteq S_n \setminus E_n$ of positive measure we have $D \notin \Gamma$. Then supp$M = S \setminus \bigcup E_n$. □

We now consider the question, whether an ideal space X contains a unity, i.e. a function $u \in X$ with supp$u$ = supp$X$. For pre-ideal spaces X we have only (cf. [48, §67 Theorem 4]):

Theorem 2.2.5. *Let X be a real-valued pre-ideal space with finite support. Then for any $\varepsilon > 0$ there exists some measurable set E with $\chi_E \in X$ and* mes(supp$X \setminus E) < \varepsilon$.

Proof. Let Γ be the set of all measurable sets $E \subseteq$ supp$X$ with $\chi_E \in X$. Choose a sequence $E_n \in \Gamma$ with

$$\text{mes}E_n \to \sup\{\text{mes}E : E \in \Gamma\} = m.$$

Let $T = \bigcup E_n$. If supp$X \setminus T$ has positive measure, there exists some $x \in X$ with mes(supp$x \setminus T) > 0$. Hence for some k the set

$$M = \{s \in \text{supp}X \setminus T : |x(s)| \geq k^{-1}\} \in \Gamma$$

must have positive measure. But then

$$\text{mes}(E_n \cup M) = \text{mes}E_n + \text{mes}M \to m + \text{mes}M,$$

which is by $E_n \cup M \in \Gamma$ a contradiction to the definition of m. Thus supp$X \setminus T$ has measure zero. Now we define $D_n = \bigcup_{k=1}^{n} E_k \in \Gamma$ and get by $D_1 \subseteq D_2 \subseteq \ldots$ and $\bigcup D_n = T$ that

$$\lim \text{mes}D_n = \text{mes}T = \text{mes}(\text{supp}X).$$

Hence for n large enough we have mes(supp$X \setminus D_n) < \varepsilon$. □

Corollary 2.2.7. *Let X be a real-valued pre-ideal space with σ-finite* supp$X$ = $\bigcup S_n$, *where $S_1 \subseteq S_2 \subseteq \ldots$ and* mes$S_n < \infty$. *Then there exists a sequence $E_n \subseteq S_n$, $E_1 \subseteq E_2 \subseteq \ldots$ with $\bigcup E_n$ =* supp$X$, *such that $\chi_{E_n} \in X$.*

Proof. By Theorem 2.2.5 there exist $D_{nm} \subseteq S_n$ with $\text{mes}(S_n \setminus D_{nm}) < m^{-1}$ and $\chi_{D_{nm}} \in X$. The measure of $F_n = S_n \setminus \bigcup_m D_{nm}$ is less than any m^{-1}, whence zero. Now let $E_k = \bigcup_{n,m=1}^{k}(D_{nm} \cup F_n)$. □

For ideal spaces we can say more (see also [50]):

Theorem 2.2.6. *Let X be an ideal space, and E be a σ-finite subset of* supp$X$. *Then there exists some $u \in X$ with* supp$u = E$.

Proof. Let X_E be the restriction of X to functions vanishing outside E. Lemma 2.2.8 implies supp$X_E = E$. Hence, by Corollary 2.2.7 there exists a sequence of functions $x_n \in X$ such that $\bigcup \text{supp} x_n = E$. Now apply Corollary 2.2.4, observing that X_E is an ideal space by Corollary 2.2.2. □

Theorem 2.2.6 does not hold for pre-ideal spaces, even if the underlying measure space is finite:

Example 2.2.3. Let $S = [0, 1]$, and X be the subspace of $L_1([0, 1])$ of functions vanishing on some interval $[0, \delta)$, $\delta > 0$. Then supp$X = S$, but by definition there is no $u \in X$ with supp$u = S$.

3. Ideal Spaces with Additional Properties

3.1 The W-Property

Definition 3.1.1. *A metric linear space X of (classes of) measurable functions has the* W-property, *if $x_n \to x$ in X implies $x_n \to x$ in measure.*

The W-property (W=weighting) means for a pre-ideal space that the norm weighs all parts of the underlying measure space with a least minimal amount:

Lemma 3.1.1. *A pre-ideal space X has the W-property, if and only if there exists a monotonically increasing function $c : [0, \infty] \to [0, 1]$, $c(m) > 0$ for $m > 0$, such that (for the real form of X)*

$$||\chi_E|| \geq c(\text{mes}E) \text{ for each } E \text{ with } \chi_E \in X. \tag{3.1}$$

Proof. 1. Let X have the W-property. Define

$$c(m) = \inf\{1, ||\chi_E|| : \chi_E \in X \text{ and } \text{mes}E \geq m\}.$$

c is monotonically increasing by definition. If $c(m) = 0$ for some $m > 0$, there exists a sequence E_n with $||\chi_{E_n}|| \to 0$ and $\text{mes}E_n \geq m$, i.e. $x_n = \chi_{E_n}$ tends to zero in norm, but not in measure.
2. Let such a c be given, and $||x_n - x|| \to 0$. Fix $\varepsilon > 0$. Let $M_n = \{s : |x_n(s) - x(s)| \geq \varepsilon\}$. By (3.1)

$$||x_n - x|| \geq ||\varepsilon\chi_{M_n}|| \geq \varepsilon c(\text{mes}M_n).$$

Hence $||x_n - x|| \to 0$ implies $\text{mes}M_n \to 0$ by the monotonicity and positivity of c. □

Ideal spaces X over finite measure spaces have the W-property [50]: Indeed, the W-property means that the natural embedding of X into the space of measurable functions is continuous. By the closed graph theorem (for F-spaces, [15, Theorem II.2.4]) this is equivalent to the fact that $x_n \to x$ in X and $x_n \to y$ in measure implies $x = y$. But this is the statement of Theorem 2.2.3. (Here we needed the finiteness of the measure for the fact that the space of measurable functions is an F-space). We will show now that this result holds also, if X is not complete:

Lemma 3.1.2. *Let X be a real-valued pre-ideal space. Then for any sequence of sets E_n satisfying* $\operatorname{mes} \bigcup E_n < \infty$ *and* $\chi_{E_n} \in X$*, we have that* $\|\chi_{E_n}\| \to 0$ *implies* $\operatorname{mes} E_n \to 0$.

Proof. Otherwise there exist $\varepsilon > 0$ and E_n with $\|\chi_{E_n}\| \leq n^{-2}$ and $\operatorname{mes} E_n \geq \varepsilon$, where

$$D_n = \bigcup_{k=n}^{\infty} E_k$$

have finite measure. Thus there exist $m_n \geq n$, such that

$$F_n = \bigcup_{k=n}^{m_n} E_k$$

satisfies $\operatorname{mes}(D_n \setminus F_n) < 2^{-n}\varepsilon$. Putting $F = \bigcap F_n$ we have

$$D := \bigcap_{n=1}^{\infty} D_n \subseteq F \cup \bigcup_{n=1}^{\infty} (D_n \setminus F_n). \tag{3.2}$$

Since for any n

$$\|\chi_F\| \leq \|\chi_{F_n}\| \leq \|\sum_{k=n}^{m_n} \chi_{E_k}\| \leq \sum_{k=n}^{m_n} k^{-2},$$

we have $\|\chi_F\| = 0$, i.e. $\operatorname{mes} F = 0$. Thus (3.2) implies $\operatorname{mes} D < \operatorname{mes} F + \sum 2^{-n}\varepsilon = \varepsilon$. On the other hand, $C_n = \bigcap_{k=1}^{n} D_k$ satisfies $C_1 \supseteq C_2 \supseteq \ldots$ and $\operatorname{mes} C_1 < \infty$, hence $\operatorname{mes} D = \lim \operatorname{mes} C_n \geq \varepsilon$, a contradiction. □

We now are able to prove the important

Theorem 3.1.1. *Let X be a pre-ideal space, $x_n \to x$ in X. Then $x_n \to x$ in measure on each set of finite measure.*

Proof. Let E have finite measure, $\delta > 0$, and $E_n = \{s \in E : |x(s) - x_n(s)| \geq \delta\}$. We have to prove that $\operatorname{mes} E_n \to 0$. But since $|\delta \chi_{E_n}| \leq |x - x_n|$ implies $\chi_{E_n} \in X$ and $\|\chi_{E_n}\| \to 0$, this is a consequence of Lemma 3.1.2. □

The following corollary is due to Zabrejko [50, Theorem 3 and Lemma 2]. But our proofs of Theorem 3.1.1 and Lemma 3.1.2 are simpler (although they follow the same line).

Corollary 3.1.1. *Each pre-ideal space with finite support has the W-property.*

The corollary is somewhat surprising, if you observe that if X is a pre-ideal space over some measure space (S, Σ, μ), then X depends only on Σ, not on any values of μ (except for the null sets). Thus you may renorm μ arbitrary (without changing the null sets), and still have the same pre-ideal space. The corollary states that *any* such renorming automatically satisfies a connection between X and μ, which is expressed by Lemma 3.1.1 (as long as μ remains finite).

We remark that the previous results may fail for Y-valued pre-ideal* spaces, if Y is infinite-dimensional:

Example 3.1.1. Let $Y = L_\infty([0,1])$, $S = \{0\}$ (mes$S = 1$), and let the pre-ideal* space X consist of all functions $x : S \to Y$ with norm $\|x\| = \int_0^1 |x(0)(s)|\, ds$. Let $x_n \in X$ be defined by $x_n(0) = \chi_{[0,1/n]}$. Then $\|x_n\| \to 0$, but $x_n \not\to 0$ in measure.

In a similar way for any infinite-dimensional Y and any measure space S we may even construct counterexamples of Y-valued ideal* spaces over S, in which the above results fail, provided we may use the axiom of choice:

Example 3.1.2. The axiom of choice implies that Y has an algebraical (Hamel) base e_α, $\alpha \in A$. Without loss of generality let its elements be normed, $|e_\alpha| = 1$. The index set A contains a countable subset, without loss of generality $\mathbb{N} \subseteq A$. Now define a linear isomorphism $I : Y \to Y$ by the equalities $I(e_\alpha) = e_\alpha$ for $\alpha \notin \mathbb{N}$ and $I(e_n) = n^{-1} e_n$. Let a new norm on Y be given by $|y|^* = |Iy|$. Since by definition I is an isometric isomorphism of $(Y, |\cdot|)$ and $(Y, |\cdot|^*)$, the space Y must also be complete with respect to the new norm $|\cdot|^*$. Hence, if X consists of all measurable functions $x : S \to Y$ with finite norm $\|x\| = \operatorname{ess\,sup} |x(s)|^*$, X is a Y-valued ideal* space over S. However, the sequence $x_n(s) \equiv e_n$ satisfies $\|x_n\| = n^{-1} \to 0$ but $|x_n(s)| = 1$ for any s.

The axiom of choice is essential in Example 3.1.2: If Y is a Banach space with norm $|\cdot|$, and $|\cdot|^*$ is another norm on Y, then the pre-ideal* space defined by $\|x\| = \operatorname{ess\,sup} |x(s)|^*$ is an ideal* space, if and only if Y is also complete with respect to the norm $|\cdot|^*$. Now consider the identity map $I : (Y, |\cdot|^*) \to (Y, |\cdot|)$. Without the axiom of choice or a similar axiom, it may not be disproved that I is bounded (see [17] or [47, Theorem 2.2]). But if I is bounded, there exists some $C > 0$ satisfying $|y| \le C\,|y|^*$ for any $y \in Y$, hence $\|x_n\| \to 0$ implies $\operatorname{ess\,sup} |x_n(s)| \to 0$, in particular $x_n \to 0$ in measure.

We now have a sharpening of Lemma 2.2.1 in σ-finite measure spaces:

Corollary 3.1.2. *Let X be a pre-ideal space. If $x_n \to x$ in X, and if all* suppx_n *and* supp$x$ *are σ-finite, then there exists a subsequence with $x_{n_k} \to x$ a.e.*

Proof. By Theorem 3.1.1 we have $x_n \to x$ extended (see Definition 2.2.3). Now use Lemma 2.2.3, and recall that x_n and x are finite a.e. by Proposition 2.1.1. □

The following consequence of Corollary 3.1.1 will be of particular use for the integration of functions with values in ideal spaces (Theorem 4.4.1).

Lemma 3.1.3. *Let T be a finite measure space, S be a σ-finite measure space, and U be a pre-ideal space over S with finite support. If x_n is measurable on $T \times S$, such that $y_n(t) := \|x_n(t,\cdot)\|_U \to 0$ in measure, then $x_n \to 0$ in the product-measure.*

Proof. Let c be the function of Lemma 3.1.1. Let $\varepsilon > 0$, $M_n = \{(t,s) : |x_n(t,s)| \geq \varepsilon\}$. By (3.1)

$$y_n(t) \geq \|\varepsilon \chi_{M_n}(t,\cdot)\|_U \geq \varepsilon c(\text{mes}\{s : (t,s) \in M_n\}) = \varepsilon c(\int_S \chi_{M_n}(t,s)ds).$$

Hence $c(\int_S \chi_{M_n}(\cdot,s)ds) \to 0$ in measure, i.e. for any $\delta > 0$ we have by $c(\delta) > 0$:

$$\text{mes}\{t : c(\int_S \chi_{M_n}(t,s)ds) \geq c(\delta)\} \to 0,$$

which implies by the monotonicity of c

$$\text{mes}\{t : \int_S \chi_{M_n}(t,s)ds \geq \delta\} \to 0,$$

i.e. $\int_S \chi_{M_n}(\cdot,s)ds \to 0$ in measure. By Lebesgue's dominated convergence theorem (with dominating constant function $\int_{\text{supp}U} ds = \text{mes}(\text{supp}U)$) we have

$$\int_T \int_S \chi_{M_n}(t,s)dsdt \to 0.$$

But this means $\text{mes}M_n \to 0$ by Fubini-Tonelli. □

Another form of the W-property is worth mentioning:

Lemma 3.1.4. *A normed linear space of measurable functions has the W-property, if and only if it has the following property: Whenever $B \subseteq X$ is bounded in norm, then*

$$\lim_{n\to\infty} \sup_{x\in B} \text{mes}\{s : |x(s)| \geq n\} = 0,$$

i.e. B is bounded in measure.

Proof. If X does not have this property, there exists a bounded sequence $x_n \in X$ and $\varepsilon > 0$ with $\operatorname{mes}\{s : |x_n(s)| \geq n\} \geq \varepsilon$ for all n. Then $y_n = n^{-1}x_n$ converges to zero in norm, but not in measure.
Conversely, if X does not have the W-property, then there exists a sequence $x_k \in X$ with $\|x_k\| \to 0$ but $x_k \not\to 0$ in measure. Thus there exist $\delta > 0$, $\varepsilon > 0$ such that $\operatorname{mes}\{s : |x_k(s)| \geq \delta\} \geq \varepsilon$ for infinitely many x_k, without loss of generality for all. Let B consist of all $\|x_k\|^{-1} x_k$. Given some n, choose k with $\|x_k\| \leq n^{-1}\delta$. Then

$$\sup_{x \in B} \operatorname{mes}\{s : |x(s)| \geq n\} \geq \operatorname{mes}\{s : \left| \|x_k\|^{-1} x_k(s) \right| \geq n\} \geq \varepsilon.$$

□

In this sense Corollary 3.1.1 implies (cf. also [50, Theorem 4]):

Corollary 3.1.3. *Let X be a pre-ideal space, and y be measurable, such that the measure of* $\operatorname{supp} y$ *is finite. Then for any bounded $B \subseteq X$ we have*

$$\lim_{n \to \infty} \sup_{x \in B} \operatorname{mes}\{s \in \operatorname{supp} y : |x(s)| \geq n\,|y(s)|\} = 0.$$

Proof. Let $M = \operatorname{supp} y$, and define for $x \in X$

$$Zx(s) = \begin{cases} |y(s)|^{-1} x(s) & \text{if } s \in M, \\ 0 & \text{if } s \notin M. \end{cases}$$

The set $X(|y|)$ of all Zx becomes a pre-ideal space with norm $\|Zx\|_{X(|y|)} = \|P_M x\|$, where $P_M x(s) = \chi_M(s)x(s)$. By Corollary 3.1.1, $X(|y|)$ has the W-property. Lemma 3.1.4 applied to $X(|y|)$ now yields

$$\lim_{n \to \infty} \sup_{x \in B} \operatorname{mes}\{s : |Zx(s)| \geq n\} = 0,$$

which implies the statement. □

All previous results may fail, if the conditions of finite measure are dropped, even in ideal spaces with σ-finite support:

Example 3.1.3. Let X be the ideal space over $[0, \infty)$, defined by the norm

$$\|x\| = \int_0^\infty |x(s)|\, e^{-s} ds.$$

Then $E_n = [n, n+1]$, $x_n = \chi_{E_n}$ resp. $x_n(t,s) = \chi_{E_n}(s)$ are counterexamples for Lemma 3.1.1/3.1.2, Theorem 3.1.1 and Corollary 3.1.1, resp. Lemma 3.1.3. Lemma 3.1.4 even fails for $B = \{id\}$, where the mentioned limit is not even real. Corollary 3.1.3 fails, if $\operatorname{supp} y$ is not finite, even if X has the W-property, as we see for $X = L_1([1, \infty))$, $y(s) = s^{-3}$, $B = \{s \mapsto s^{-2}\}$.

To avoid such difficulties, one often replaces the measure on the measure space (S, σ, μ) by a normalized one (S, σ, ν), i.e. by a measure with the same null sets but $\nu(S) = 1$. If μ is σ-finite, you can always do this by Corollary 2.2.6.

3.2 Completeness and Perfectness

Now we want to study conditions, which ensure that a given pre-ideal space is in fact an ideal space, i.e. that it is complete. That this is usually not the case shows Example 2.2.3. This example is very simple, since the space given there contains no function with full support (hence Theorem 2.2.6 implies that the space may not be complete, but this can of course also be checked straightforwardly). Also the completion of the space in this example is (isomorphic to) the ideal space $L_1([0,1])$. An example of a noncomplete pre-ideal space, which contains functions with full support and whose completion is not (isomorphic to) an ideal space is given in [48, §64]:

Example 3.2.1. Let X be the pre-ideal space of sequences $x = (x_k)_k$ with finite norm

$$\|x\| = \sum_{k=1}^{\infty} 2^{-k} |x_k| + \limsup_{k\to\infty} |x_k| .$$

Define the sequence $x_n = (x_k^n)_k \in X$ by

$$x_k^n = \begin{cases} 1 & k \le n, \\ 0 & k > n. \end{cases}$$

Then $\|x_{n+m} - x_n\| = \sum_{k=n+1}^{n+m} 2^{-k}$ implies that x_n is a Cauchy sequence in X. Assume, X is a subspace of some ideal space Y. Then x_n must converge in norm to some $x \in Y$. By Theorem 2.2.3 the limit x must be the pointwise limit, i.e. $x = (1)_k$. But then $\|x - x_n\| \to 1$, a contradiction.

Now we are going to characterize ideal spaces. Our main goal in this context will be Theorem 3.2.1, which states that completeness of a pre-ideal space is equivalent to the following property:

Definition 3.2.1. *A pre-ideal* space X has the* Riesz-Fischer property, *if for any sequence $x_n \in X$ with $\sum_{n=1}^{\infty} \|x_n\| < \infty$ we have that $\sum_{n=1}^{\infty} x_n$ converges extended* to some function in X.*

Sometimes in literature additionally the estimate $\|\sum x_n\| \le \sum \|x_n\|$ is assumed to hold, see e.g. [50]. We will see that for pre-ideal spaces this inequality is a consequence of the Riesz-Fischer property (Corollary 3.2.2). For a straightforward proof of this fact see [48, §64 Theorem 1].

For a pre-ideal space the previous definition depends only on its real form. And for the real form the definition coincides with the one given in [48, §64]:

Lemma 3.2.1. *A Y-valued pre-ideal space X_Y has the Riesz-Fischer property, if and only if its real form X has the following property: If $0 \le x_n \in X$ satisfies $\sum_{n=1}^{\infty} \|x_n\| < \infty$, then $\sum_{n=1}^{\infty} x_n$ converges a.e. to some function in X.*

Proof. We first prove necessity: Choose $e \in Y$ with $|e| = 1$. Let $0 \le x_n \in X$ satisfy $\sum \|x_n\| < \infty$. Now define $y_n = ex_n$. By assumption $\sum y_n = e \sum x_n$ converges extended* to some $y \in X_Y$. Lemma 2.2.6 implies $y = ex$ for some $x \in X$. Hence $\sum x_n$ converges extended* to some $x \in X$. Corollary 2.2.1 implies by $x_n \ge 0$ even that $\sum x_n$ converges to x a.e. in the extended sense, hence a.e.
We now prove sufficiency: Let $x_n \in X_Y$ satisfy $\sum \|x_n\| < \infty$. Since $|x_n| \in X$, we have by assumption that $\sum |x_n|$ converges a.e. to some $y \in X$. Thus $\sum x_n$ converges a.e. to some x, which satisfies $|x| \le y$, whence $x \in X_Y$. □

In the sufficient part we have even seen that $\sum x_n$ converges a.e. to x. This implies that for pre-ideal spaces we may replace 'extended* convergence' by 'convergence a.e.' in Definition 3.2.1:

Lemma 3.2.2. *A pre-ideal space X has the Riesz-Fischer property, if and only if for any sequence $x_n \in X$ with $\sum_{n=1}^{\infty} \|x_n\| < \infty$ we have that $\sum_{n=1}^{\infty} x_n$ converges a.e. to some function in X.*

To check the Riesz-Fischer property the following lemma is very useful (this is the characterization used in [38]):

Lemma 3.2.3. *A pre-ideal space has the Riesz-Fischer property, if and only if its real form X has the following property: If $x_n \ge 0$ is a monotonically increasing Cauchy-sequence in X, then* $\sup x_n \in X$.

Proof. We use Lemma 3.2.1.
For sufficiency, let $0 \le x_n \in X$ and $\sum \|x_n\| < \infty$. Then $y_k = \sum_{n=1}^{k} x_n \ge 0$ is a monotonically increasing Cauchy-sequence, since

$$\|y_{k+m} - y_k\| \le \sum_{n=k+1}^{k+m} \|x_n\| .$$

Thus $x = \sup y_k \in X$. Proposition 2.1.1 now implies that $\sum x_n = x$ converges a.e.
For necessity, let $0 \le x_n \in X$ be a monotonically increasing Cauchy-sequence, and X have the Riesz-Fischer property. Choose a subsequence x_{n_k} with $\sum_{k=1}^{\infty} \|x_{n_{k+1}} - x_{n_k}\| < \infty$, and define $y_k = x_{n_{k+1}} - x_{n_k}$. Then $\sum_{n=1}^{\infty} \|y_n\| < \infty$, thus by assumption the series $y = \sum_{k=1}^{\infty} y_k$ converges a.e. with $y \in X$. But the partial sums of this series, $\sum_{k=1}^{m} y_k = x_{n_{m+1}} - x_{n_1}$, converge by the monotony of x_n to $y = (\sup x_n) - x_{n_1}$. Thus $\sup x_n = y + x_{n_1} \in X$. □

Now we will prove the mentioned Theorem 3.2.1 about completeness of a pre-ideal space. For real-valued spaces (over finite measure spaces) it is proved in [38, Theorem 1.5] (see also [48, §64 Theorem 2]). We will modify the sufficient part of the proof to hold on arbitrary pre-ideal spaces, and for the necessary part our very general Theorem 2.2.3 even will allow us to consider arbitrary measure spaces. We state one part of the proof as a lemma for later reference:

Lemma 3.2.4. *If a pre-ideal space X has the Riesz-Fischer property, then any Cauchy-sequence of X contains a subsequence which converges a.e. to some element of X. Furthermore, this subsequence x_n may be chosen such that $|x_n| \leq |y|$ for some $y \in X$.*

Proof. Choose the subsequence x_n, such that $\sum_{n=1}^{\infty} \|x_{n+1} - x_n\| < \infty$. By Lemma 3.2.1, applied to $y_n = |x_{n+1} - x_n|$, we see that

$$w = \sum_{k=1}^{\infty} |x_{k+1} - x_k|$$

converges a.e., where w belongs to the real form of X. Analogously, by Lemma 3.2.2 the series

$$x = \sum_{k=1}^{\infty} (x_{k+1} - x_k) \tag{3.3}$$

converges a.e. with $x \in X$. But the partial sums of (3.3) are just

$$\sum_{k=1}^{n-1} (x_{k+1} - x_k) = x_n - x_1, \tag{3.4}$$

i.e. x_n converges a.e. to $x + x_1 \in X$. Furthermore, (3.4) implies $|x_n - x_1| \leq w$ and thus the additional statement by $|x_n| \leq |x_n - x_1| + |x_1| \leq w + |x_1|$. □

Theorem 3.2.1. *A pre-ideal space X is an ideal space, if and only if it has the Riesz-Fischer property.*

Proof. Let X have the Riesz-Fischer property, and $x_n \in X$ be Cauchy. Since it suffices to prove that x_n contains a convergent subsequence, we may assume by Lemma 3.2.4 that x_n converges almost everywhere to some $x \in X$, and $\|x_{n+1} - x_n\| \leq n^{-3}$. Define

$$y_n = \sum_{k=1}^{n} k\,|x_{k+1} - x_k|\,.$$

By the Riesz-Fischer property the supremum of y_n,

$$y = \sum_{k=1}^{\infty} k\,|x_{k+1} - x_k|\,,$$

belongs to X. Thus by

$$n\,|x - x_n| = |n \sum_{k=n}^{\infty} (x_{k+1} - x_k)| \leq \sum_{k=n}^{\infty} k\,|x_{k+1} - x_k| \leq y$$

we have $n\,\|x - x_n\| \leq \|y\|$, hence $\|x - x_n\| \leq \|y\|/n \to 0$.
The converse follows by Lemma 3.2.3, using Lemma 2.2.7 and Corollary 2.2.3, since each monotonically increasing sequence converges to its supremum a.e. in the extended sense. □

The theorem implies in view of Lemma 3.2.1 that a pre-ideal space is an ideal space, if and only if its real form is an ideal space.

In ideal spaces we can strengthen Lemma 2.2.1 now:

Corollary 3.2.1. *Let X be an ideal space, $x_n \to x$ in X. Then there exists a subsequence and some $y \in X$, such that $x_{n_k} \to x$ a.e. and $|x_{n_k}| \leq |y|$.*

Proof. By Theorem 3.2.1, X has the Riesz-Fischer property, thus by Lemma 3.2.4 some subsequence x_{n_k} converges to some w a.e. and is dominated by some $y \in X$. Corollary 2.2.3 implies $w = x$. □

In contrast to Corollary 3.1.2, the above Corollary 3.2.1 does not need the condition that some sets are σ-finite, and we additionally get a dominating function $y \in X$. Of course we have to pay the price that X must be complete. Indeed, in pre-ideal spaces, such a $y \in X$ need not exist, even if the underlying measure space has finite measure:

Example 3.2.2. Let $S = [0, 1]$, and X be the subspace of $L_1(S)$ of functions vanishing a.e. on some interval $[0, \delta)$, $\delta > 0$. The sequence $x_n = \chi_{[n^{-1}, 2n^{-1}]}$ converges to $x = 0$ in X, but of course no subsequence is dominated by a function $y \in X$.

As another consequence we have that for pre-ideal spaces the Riesz-Fischer property is equivalent to an apparently more restrictive property. Sometimes this property is called Riesz-Fischer property in literature (see e.g. [50]):

Corollary 3.2.2. *A pre-ideal space X has the Riesz-Fischer property, if and only if for any sequence $x_n \in X$ with $B = \sum \|x_n\| < \infty$ we have that $x = \sum x_n$ converges a.e. with $x \in X$, and additionally the estimate $\|x\| \leq B$ holds.*

Proof. Sufficiency is evident. For necessity by Lemma 3.2.2 it suffices to prove the norm estimate. Since $y_n = \sum_{k=1}^{n} x_k$ is a Cauchy-sequence in X, Theorem 3.2.1 and Corollary 2.2.3 imply $\|x - y_n\| \to 0$, hence $\|y_n\| \to \|x\|$. Thus by $\|y_n\| \leq B$ we have $\|x\| \leq B$. □

Definition 3.2.2. *Let X be a pre-ideal* space. X is called* semi-perfect, *if the fact that a sequence $x_n \in X$ with bounded norms converges to some x a.e. in the extended sense implies $x \in X$.*

X is called almost α-perfect, *if for any sequence $x_n \in X$, converging a.e. to some $x \in X$, one has*

$$\|x\| \le \alpha \liminf_{n\to\infty} \|x_n\| . \tag{3.5}$$

If X is both, semi-perfect and almost α-perfect, it is called α-perfect.
In the case $\alpha = 1$, we just say that X is perfect *resp.* almost perfect.

If a pre-ideal* space is almost α-perfect, there exists a minimal such α. This follows by the fact that (3.5) also holds for $\alpha = \beta$, if it holds for any $\alpha > \beta$. The choice $x_n = x$ shows $\alpha \ge 1$. We will see that for pre-ideal spaces over σ-finite measure spaces we may pass to an equivalent norm to have $\alpha = 1$ (Corollary 3.2.8).

To check (3.5) it suffices, of course, just to consider sequences x_n, which are additionally bounded in norm. Even more, it suffices to consider sequences, for which $\lim \|x_n\| = \liminf \|x_n\|$ exists and is finite (since for any sequence with bounded norms, we may find such a subsequence).

Equivalently, we could have chosen 'extended convergence' in Definition 3.2.2:

Lemma 3.2.5. *Let X be a pre-ideal* space.*

1. *X is semi-perfect, if and only if we have $x \in X$, whenever a sequence $x_n \in X$ with bounded norms converges extended to x.*
2. *X is almost α-perfect, if and only if (3.5) holds for any sequence $x_n \in X$ converging extended to $x \in X$.*

Proof. If $x_n \in X$ converges extended to x, choose a subsequence with $\|x_{n_k}\| \to \liminf \|x_n\|$. Now observe that by Lemma 2.2.3 we may additionally assume that $x_{n_k} \to x$ a.e. in the extended sense. □

We emphasize that in Definition 3.2.2 we speak of convergence a.e. in the extended sense (and not just of convergence a.e.). This means that we do not assume a priori that x is finite a.e., but have this as a consequence. This becomes important in Corollary 3.2.4 to ensure that semi-perfect pre-ideal spaces have the Riesz-Fischer property. However, if the underlying measure space has the finite subset property, we may assume a priori that x is finite a.e.:

Lemma 3.2.6. *Let X be a pre-ideal space, such that* supp$x$ *has the finite subset property for any $x \in X$. Then X is semi-perfect, if and only if for any bounded sequence $x_n \in X$ with $x_n \to x$ a.e. we have $x \in X$.*

Proof. If $x_n \in X$ is bounded with $x_n \to x$ a.e. in the extended sense, then x is finite a.e. by Lemma 2.2.2, hence $x_n \to x$ a.e. □

Hence for pre-ideal spaces over finite measure spaces our definition of perfectness coincides with the one given in [50]:

Corollary 3.2.3. *Let X be a pre-ideal space with finite support. Then a pre-ideal space X is perfect, if and only if its closed unit ball B is closed with respect to convergence in measure.*

Proof. Let B be closed in measure, and the bounded sequence $x_n \in X$ converge a.e. (whence in measure) to some x. Now observe that by assumption $\|x_n\| \leq C$ implies $x \in X$ and $\|x\| \leq C$.
Conversely, let X be perfect. Let $x_n \in B$ converge in measure to some x. Since a subsequence converges a.e., we have $x \in B$. □

All properties of Definition 3.2.2 depend only on the real form of X. Here even more, it suffices to consider monotonically increasing nonnegative sequences (recall that we do not distinguish in the notation between X and the real form of X).

Lemma 3.2.7. *Let X be a Y-valued pre-ideal space.*

1. *X is semi-perfect, if and only if for any monotonically increasing sequence of nonnegative functions $x_n \in X$ with bounded norms we have $\sup x_n \in X$.*
2. *X is almost α-perfect, if and only if* (3.5) *holds for any monotonically increasing sequence of nonnegative functions $x_n \in X$ with $x = \sup x_n \in X$.*

Proof. We first prove sufficiency: Let $x_n \to x$ a.e. in the extended sense, where $x_n \in X$ has bounded norms. Then $y_n = \inf_{k \geq n} |x_k|$ is monotonically increasing and nonnegative with $y = \sup y_n = \liminf |x_n| = \lim |x_n| = |x|$. Thus $y \in X$ implies $x \in X$, and $\|y\| \leq \alpha \liminf \|y_n\|$ implies (3.5) by $\|y_n\| \leq \|x_n\|$.
For necessity, let $x_n \in X$ be nonnegative and monotonically increasing with bounded norms, $x = \sup x_n$. Then, for $e \in Y$, $|e| = 1$ we have $x_n e \to xe$ a.e. in the extended sense. If X is semi-perfect, we have $xe \in X$, hence $x \in X$. If X is almost α-perfect and $x \in X$, we have $\|xe\| \leq \alpha \liminf \|x_n e\|$, thus (3.5) holds. □

In the terminology of [48] the previous lemma means that a pre-ideal space is (semi-)perfect, if and only if its real form has the (weak) Fatou property. As a first consequence we have by Theorem 3.2.1:

Corollary 3.2.4. *Each semi-perfect pre-ideal space is an ideal space.*

Proof. A semi-perfect pre-ideal space has the Riesz-Fischer property by Lemma 3.2.3, since any Cauchy-sequence is bounded in norm. □

As another consequence we have that almost perfect pre-ideal spaces have some weak Levi property:

Corollary 3.2.5. *Let X be an almost α-perfect pre-ideal space, $x \in X$, and x_n be a monotonically increasing sequence of nonnegative measurable functions, converging to $|x|$ a.e. Then*

$$\alpha^{-1}\,\|x\| \le \lim_{n\to\infty} \|x_n\| \le \|x\|\,.$$

In particular, if X is almost perfect, we have $\|x_n\| \to \|x\|$.

Proof. Since $\|x_1\| \le \|x_2\| \le \ldots \le \|x\|$ the limit exists and is not bigger than $\|x\|$. (3.5) implies the other estimate. □

Perfectness is equivalent to the fact that the real form of the space satisfies Fatou's lemma (cf. also [48, §65 Theorem 3]):

Lemma 3.2.8. *Let X be a Y-valued pre-ideal space.*

1. *X is semi-perfect, if and only if for any sequence of nonnegative functions $x_n \in X$, which is bounded in norm, we have $\liminf x_n \in X$.*
2. *X is almost α-perfect, if and only if (3.5) holds for any sequence of nonnegative functions $x_n \in X$ with $x = \liminf x_n \in X$.*

Proof. The sufficiency follows immediately by Lemma 3.2.7. But this implies also necessity: Indeed, let $y_n = \inf_{n\ge k} x_k$. Then $y_n \ge 0$ is monotonically increasing with $\liminf y_n = \sup y_n = \liminf x_n = x$. Now observe that $\|y_n\| \le \|x_n\|$, and apply Lemma 3.2.7. □

We remark that a semi-perfect ideal space need not be almost perfect (see [48, §65 Exercise 1]):

Example 3.2.3. Let $\beta \ge 0$, and X consist of all sequences $x = (x_k)_k$ with finite norm

$$\|x\| = \sup_k |x_k| + \beta \limsup_k |x_k|\,.$$

Then X is α-perfect with $\alpha = 1+\beta$ (and an ideal space by Corollary 3.2.4). To see this, we use Lemma 3.2.7: Let $x_n = (x_k^n)_k$ be monotonically increasing, nonnegative and bounded in norm: $\|x_n\| \le C$. Let $y = (y_k)_k$ be the supremum of x_n. Then trivially $|y_k| \le C$, hence $\|y\| \le C + \beta C = (1+\beta)C < \infty$, and X is α-perfect with $\alpha = 1+\beta$.

But X is not almost α-perfect for $\alpha < 1+\beta$: Define a sequence $x_n = (x_k^n)_k$ by

$$x_k^n = \begin{cases} 1 & k \le n \\ 0 & k > n. \end{cases}$$

Then $\|x_n\| = 1$, but the limit function x satisfies $\|x\| = 1+\beta$.

However, a semi-perfect pre-ideal space is always almost α-perfect. The main part of the following proof is due to [2] (see [48, §65 Theorem 2]):

Theorem 3.2.2. *A pre-ideal space is semi-perfect, if and only if it is α-perfect for some $\alpha > 0$.*

Proof. Assume, X is semi-perfect, but not almost α-perfect for any $\alpha > 0$. Then, by Lemma 3.2.7 for any k there exists a sequence of nonnegative functions $x_n^k \in X$, which converges monotonically increasing to $x^k \in X$ with

$$\|x^k\| > k^3 \lim_{n\to\infty} \|x_n^k\| = k^3 c_k.$$

By $x^k \neq 0$ we have $c_k > 0$. Since x_n^k is monotonically increasing we have $\|x_n^k\| \leq c_k$. Thus, the sequence $y_n = \sum_{k=1}^n k^{-2} c_k^{-1} x_n^k$ is monotonically increasing and has bounded norm:

$$\|y_n\| \leq \sum_{k=1}^{\infty} k^{-2} c_k^{-1} \|x_n^k\| \leq \sum_{k=1}^{\infty} k^{-2}.$$

Lemma 3.2.7 now implies $y = \sup y_n \in X$, but on the other hand we have by $y \geq k^{-2} c_k^{-1} x^k$ that

$$\|y\| > k^{-2} c_k^{-1} k^3 c_k = k$$

for any k, which is not possible. □

It is easy to find an example of a pre-ideal space, which is not almost α-perfect for any $\alpha > 0$. However, it is not easy to find an ideal space with such a property. Thus you might suspect that Theorem 3.2.2 may be sharpened (see Corollary 3.2.4) in the sense that any ideal space is almost α-perfect. It turns out that this is not true:
There exists an ideal space, which is not almost α-perfect for any $\alpha > 0$ (and also not semi-perfect by Theorem 3.2.2).

Example 3.2.4. Let X be the pre-ideal space of double-sequences $x = (x_{km})_{km}$ with finite norm

$$\|x\| = \sup_{k,m} |x_{km}| + \sup_k \limsup_m k\, |x_{mk}|\,.$$

At first we show that X is an ideal space. For this purpose we use Theorem 3.2.1. Thus, let $x_n = (x_{mk}^n)_{mk}$ be a monotonically increasing nonnegative Cauchy-sequence in X. Then x_{mk}^n converges uniformly to some x_{mk}. We have to prove that $x = (x_{mk})_{mk}$ belongs to X. Since $\|x_n\| \leq C$ for some $C > 0$, we have for any k

$$\limsup_m |x_{mk}^n| \leq k^{-1} C.$$

But since $x_{mk}^n \to x_{mk}$ uniformly, this also implies

$$\limsup_m |x_{mk}| \leq k^{-1}C.$$

But this in turn implies that

$$\sup_{k,m} |x_{km}| + \sup_k \limsup_m k|x_{mk}| \leq C + C < \infty,$$

whence $x \in X$, and thus X is an ideal space.
Now assume that X is almost α-perfect for some natural number α. Then define $x_n = (x^n_{mk})_{mk}$ by

$$x^n_{mk} = \begin{cases} 1 & k = \alpha \text{ and } m \leq n \\ 0 & \text{otherwise.} \end{cases}$$

We have $||x_n|| = 1$, and x_n converges a.e. to $x = (x_{mk})_{mk}$, where $x_{mk} = 0$ for $k \neq \alpha$ and $x_{m\alpha} = 1$. Thus $||x|| = 1 + \alpha$, and (3.5) is not satisfied.

The following example shows that for any $\alpha \geq 1$ there exist almost α-perfect ideal spaces, which are neither semi-perfect nor almost α-perfect for any smaller α:

Example 3.2.5. Let $\beta \geq 0$, and X be the pre-ideal space of all sequences $x = (x_k)_k$ with $x_{2k} \to 0$ and with finite norm

$$||x|| = \sup |x_k| + \beta \limsup_{k\to\infty} |x_{2k+1}|.$$

Let $x_n = (x^n_k)_k \in X$ be nonnegative and monotonically increasing with bounded norm $||x_n|| \leq C$. Then $|x^n_k| \leq C$, thus $x^n_k \to y_k$ for $n \to \infty$. Since $|y_k| \leq C$ the limit $y = (y_k)_k$ has norm $||y|| \leq C + \beta C$. On the one hand, this shows that X is almost α-perfect with $\alpha = 1 + \beta$. On the other hand, if x_n was Cauchy, we even have $x^n_k \to y_k$ uniformly in k, thus $y_{2k} \to 0$, hence $y \in X$. But this means that X has the Riesz-Fischer property, hence X is an ideal space by Theorem 3.2.1. Finally, the sequence $x_n = (x^n_k)_k$ with

$$x^n_k = \begin{cases} 1 & k \text{ is even and } k \leq n, \\ 0 & \text{otherwise,} \end{cases}$$

shows that X is not semi-perfect, and the sequence $x_n = (x^n_k)_k$ with

$$x^n_k = \begin{cases} 1 & k \text{ is odd and } k \leq n, \\ 0 & \text{otherwise,} \end{cases}$$

shows that X is not almost α-perfect for $\alpha < 1 + \beta$.

Now we address the question, in which sense a given pre-ideal* space can be changed to be (almost) perfect. The most important answer to this is given by the Lorentz seminorm:

Definition 3.2.3. *Let X be a pre-ideal* space. Then for any measurable x we define the* Lorentz seminorm $\|x\|_L$ *as the infimum of all* $\liminf \|x_n\|$, *where $x_n \in X$ converges a.e. in the extended sense to x (let $\|x\|_L = \infty$ if such a sequence does not exist). By X_L we denote the* Lorentz space *of all x with $\|x\|_L < \infty$.*

It is clear, that $\|\cdot\|_L$ is a seminorm on X_L. For the triangle inequality observe that by passing to a subsequence, we always may assume for the sequences in Definition 3.2.3 that $\|x_n\| \to \liminf \|x_n\|$.

Remark 3.2.1. If X is an almost α-perfect pre-ideal* space, the Lorentz seminorm is an equivalent norm on X: We have $\|x\| \geq \|x\|_L \geq \alpha^{-1} \|x\|$.

Remark 3.2.2. $\|x\|_L$ is the infimum of all $\liminf \|x_n\|$, where $x_n \in X$ converges extended to x.

Proof. If $x_n \to x$ extended, choose a subsequence with $\|x_{n_k}\| \to \liminf \|x_n\|$. By Lemma 2.2.3 you may assume that $x_{n_k} \to x$ a.e. in the extended sense. □

For pre-ideal spaces the definition depends only on the real form of X, and for that it coincides with the one given in [48]:

Proposition 3.2.1. *If X is a pre-ideal space, then $\|x\|_L$ is the infimum of all* $\lim \|x_n\|$, *where $x_n \geq 0$ tends monotonically increasing a.e. in the extended sense to $|x|$.*

Proof. If $x_n \to x$ a.e. in the extended sense, define $y_n = \inf_{k \geq n} |x_k|$. Then $y_n \geq 0$ tends monotonically increasing a.e. in the extended sense to $|x|$, and $|y_n| \leq |x_n|$ implies $\lim \|y_n\| \leq \liminf \|x_n\|$. □

For pre-ideal spaces over σ-finite measure spaces X_L is perfect, and the infimum in Definition 3.2.3 is in fact a minimum (see also [48, §66 Theorem 2], but we give a different proof, using Egorov's theorem):

Lemma 3.2.9. *Let X be a pre-ideal space. If $x_n \in X_L$ converges a.e. in the extended sense to x, and* $\operatorname{supp} x$ *is σ-finite, then* $\|x\|_L \leq \liminf \|x_n\|_L$.
More precisely, there exists a sequence $y_n \in X$, $|y_1| \leq |y_2| \leq \ldots$, which converges a.e. in the extended sense to x and satisfies $\lim \|y_n\| \leq \liminf \|x_n\|_L$.

Proof. Without loss of generality, let X be real-valued, and $x_n \geq 0$ tend monotonically increasing a.e. to $x \geq 0$ in the extended sense. We may even assume that each x_n is finite almost everywhere.
To any k there exists a sequence $0 \leq x_n^k \in X$, which converges monotonically increasing a.e. to x_k and satisfies

$$\|x_k\|_L \geq \lim_{n\to\infty} \|x_n^k\| - k^{-1}. \tag{3.6}$$

Let $\operatorname{supp} x = \bigcup E_j$ where $E_1 \subseteq E_2 \subseteq \ldots$ have finite measure.
By Egorov's theorem for any j, k there exists some $C_{jk} \subseteq E_j$ with $\operatorname{mes} C_{jk} < k^{-2}j^{-2}$ and $x_n^k \to x_k$ uniformly on $E_j \setminus C_{jk}$. Then $C_j = \bigcup_{m \geq j} \bigcup_k C_{mk}$ satisfies $C_1 \supseteq C_2 \supseteq \ldots$, $\operatorname{mes} C_j \to 0$, and $x_n^k \to x_k$ uniformly on each $D_j = E_j \setminus C_j$.
In particular, for each k there exists some n_k with

$$|x_n^k(s) - x_k(s)| \leq k^{-1} \qquad (s \in D_k,\ n \geq n_k),$$

and by (3.6) we may assume additionally that

$$\|x_k\|_L \geq \|x_n^k\| - 2k^{-1} \qquad (n \geq n_k).$$

Without loss of generality, we assume $n_1 \leq n_2 \leq \ldots$. Since $D_1 \subseteq D_2 \subseteq \ldots$, the monotonically increasing sequence $y_l = \inf_{k \geq l} x_{n_k}^k$ converges on each D_j a.e. in the extended sense to x. We even have $y_k \to x$ a.e. in the extended sense, since $\operatorname{supp} x \setminus \bigcup D_j = \bigcap C_j$ has measure zero. Moreover, $\|y_k\| \leq \|x_{n_k}^k\|$ implies $\liminf \|x_k\|_L \geq \lim \|y_k\|$. □

Corollary 3.2.6. *Let X be a pre-ideal space. If $x \in X_L$ has σ-finite support, then there exists a sequence $y_n \in X$ converging a.e. to x in the extended sense, satisfying $|y_1| \leq |y_2| \leq \ldots$ and $\|y_n\| \to \|x\|_L$.*

Proof. Choose y_n as in Lemma 3.2.9 for $x_n = x$. Then

$$\lim \|y_n\| \leq \liminf \|x_n\|_L = \|x\|_L .$$

Since $y_n \to x$ a.e. in the extended sense, we also have $\|x\|_L \leq \lim \|y_n\|$. □

Remark 3.2.3. The previous results about $\|\cdot\|_L$ also hold true (with the same proofs), if $\|\cdot\|$ is not a norm, but just a seminorm (i.e. $\|x\| = 0$ need not imply $x = 0$).

Now it is easy to see, that X_L is usually a pre-ideal space [48, §66 Theorem 4]:

Corollary 3.2.7. *Let X be a pre-ideal space. If* $\operatorname{supp} x$ *contains a subset E of positive finite measure, then $\|x\|_L \neq 0$.*

Proof. Without loss of generality, let $\operatorname{supp} x = E$. Then $\|x\|_L = 0$ implies in Corollary 3.2.6 that $\|y_n\| \leq \|x\|_L = 0$, i.e. $y_n = 0$, whence $x = 0$. □

Thus, over σ-finite measure spaces, the Lorentz space X_L is a perfect ideal space. Since this is obviously the smallest perfect space, in which X is continuously embedded with embedding constant ≤ 1, it might be considered as the 'perfect ideal hull' of X (see also [48] and [50]).

Combining Remark 3.2.1 and Lemma 3.2.9 we have for almost α-perfect pre-ideal spaces with σ-finite support that we may always pass to $\alpha = 1$ by just considering an equivalent norm:

Corollary 3.2.8. *If X is an almost α-perfect pre-ideal space with σ-finite support, then the equivalent Lorentz norm turns X into an almost perfect pre-ideal space.*

Finally, we give an extension of Corollary 3.2.3:

Definition 3.2.4. *A set M of measurable functions is called* normal, *if the relation $|y| \leq |x|$ a.e. for some $x \in M$ and measurable y implies $y \in M$.*

Natural examples of normal sets are pre-ideal spaces or the unit ball of pre-ideal spaces. The converse also holds under natural assumptions, as we will show now.

Definition 3.2.5. *A subset M of some linear space* does not contain a complete ray, *if*

$$\{\lambda x : \lambda \geq 0\} \not\subseteq M \qquad (x \neq 0).$$

We say that M is closed on rays, *if the set $\{\lambda \geq 0 : \lambda x \in M\}$ is closed for any x.*

Lemma 3.2.10. *A nonempty set M of measurable functions is the closed unit ball of some pre-ideal space X, if and only if M is normal, convex, closed on rays, and does not contain a complete ray.*

Proof. Necessity is obvious. Thus, assume M is normal, convex, closed on rays, and does not contain a complete ray. Let X consist of all x, for which the Minkowski functional $\|x\| = \inf\{\lambda > 0 : \lambda^{-1}x \in M\}$ is finite. To see that X is a pre-ideal space, we check only the nontrivial properties. $\|x\| = 0$ implies $x = 0$, since otherwise M would contain a complete ray $\{\lambda x : \lambda \geq 0\} \subseteq M$. Now, since M is closed on rays, we even have

$$\|x\| = \min\{\lambda > 0 : \lambda^{-1}x \in M\} \qquad (x \neq 0). \tag{3.7}$$

For the triangle inequality now just observe that $\lambda^{-1}x, \mu^{-1}y \in M$ implies

$$(\lambda + \mu)^{-1}(x + y) = \frac{\lambda}{\lambda + \mu}(\lambda^{-1}x) + \frac{\mu}{\lambda + \mu}(\mu^{-1}y) \in M$$

by the convexity of M. Finally, M coincides with $\{x \in X : \|x\| \leq 1\}$ by (3.7). □

Over finite measure spaces we can characterize now the perfect ideal spaces in terms of the metric linear space of measurable functions (cf. [15]):

Lemma 3.2.11. *A nonempty set M of measurable functions with finite support is the unit ball of some perfect ideal space X, if and only if M is normal, convex, and closed and bounded in the metric linear space of measurable functions.*

Proof. We first prove necessity. Thus, let $M = \{x \in X : \|x\| \leq 1\}$ for some perfect ideal space X. Then M is normal and convex. M is closed in measure by Corollary 3.2.3. Since M has finite support, Corollary 3.1.3 implies

$$\lim_{n\to\infty} \sup_{x\in M} \operatorname{mes}\{s : |x(s)| \geq n\} = 0,$$

which means that M is bounded in measure.
Now we prove sufficiency. Since M is bounded in measure, it may not contain a complete ray. Furthermore, M is closed on rays, since it is closed in measure: For $\lambda_n x \in M$ and $\lambda_n \to \lambda$ we have $\lambda_n x \to \lambda x$ in measure, whence $\lambda x \in M$. Thus, Lemma 3.2.10 implies that M is the closed unit ball of some pre-ideal space X. X is even a perfect ideal space by Corollaries 3.2.3 and 3.2.4. □

Sadly, both directions of Lemma 3.2.11 may fail, if supp$M$ is just σ-finite:

Example 3.2.6. Let $S = \mathbb{N}$ with the counting measure μ, and $M = \{(x_n)_n : 1 \geq |x_n| \to 0\}$. Then M is trivially normal and convex. It is easy to see that M is bounded in measure. M is also closed in measure, since convergence in μ-measure is equivalent to uniform convergence. But if M is the closed unit ball of some perfect ideal space X, just replace μ by the normalized measure

$$\nu(E) = \sum_{n\in E} 2^{-n}.$$

This does neither affect X nor M. But now $x_1 = (1, 0, \ldots)$, $x_2 = (1, 1, 0, \ldots)$, $x_3 = (1, 1, 1, 0, \ldots)$, ... converges in ν-measure to $(1, 1, \ldots) \notin M$, i.e. M is not closed in ν-measure. Thus, by Lemma 3.2.11, M may not be the closed unit ball of X, a contradiction. This shows that the sufficient part of Lemma 3.2.11 fails, if supp$M$ is not finite.

Example 3.2.7. We now show that also the necessary part may fail: Let S, μ, ν be as in the previous example, and $X = L_1(\nu)$. Then X is a perfect ideal space, but its closed unit ball $M = \{(x_n)_n : \sum 2^{-n} |x_n| \leq 1\}$ is not bounded in μ-measure, since for $x = (1, 2, \ldots) \in M$ there is no $\lambda > 0$ such that λx belongs to a sufficiently small ball around zero in the metric space of μ-measurable functions.

3.3 Regular Spaces and Convergence Theorems

The following notion is essential for calculation with ideal spaces.
Let X be a pre-ideal* space over some measure space S. For $D \subseteq S$ we write $P_D x(s) = \chi_D(s) x(s)$.

Definition 3.3.1. $x \in X$ *has* absolute continuous norm, *if*

$$\lim_{\delta \to 0} \sup_{\text{mes} D \leq \delta} \|P_D x\| = 0.$$

$x \in X$ vanishes at ∞ in norm, *if*

$$\inf_{\text{mes} E < \infty} \|P_{S \setminus E} x\| = 0.$$

The space X_0 of all $x \in X$ with absolute continuous norm is the inner-regular part *of X, the space X^0 of all $x \in X$, which vanish at ∞ in norm is the* outer-regular part *of X. The space $X_0^0 = X_0 \cap X^0$ is the* regular part *of X. If $X = X_0$ resp. $X = X^0$ we call X* inner-regular *resp.* outer-regular. *A space, which is both, inner- and outer-regular, is called* regular.
A regular perfect ideal space is called* completely regular.

Some examples: All pre-ideal* spaces over finite measure spaces are outer-regular (thus for the normalized measure, X is regular, if and only if X is inner-regular). The spaces $L_p([0,1])$, $L_p(\mathbb{R})$, l_p are regular, if and only if $p < \infty$. $L_\infty([0,1])$ is outer-regular, but not inner-regular. l_∞ is inner-regular, but not outer-regular (we assume the counting measure on $\mathbb{N}$). The regular part of l_∞ is the subspace c_0 of sequences, converging to zero. $L_\infty(\mathbb{R})$ is neither inner- nor outer-regular. The elements of $L_\infty(\mathbb{R})$, which vanish at ∞ in norm, are exactly those elements x, for which $x(t)$ tends essentially to zero for $|t| \to \infty$.

The previous example of c_0 shows that there exist regular ideal spaces, which are not completely regular (i.e. they need not be semi-perfect; observe that they are always almost perfect by Corollary 3.3.4):
In fact, c_0 is a closed subspace of l_∞ by Theorem 3.3.2, whence an ideal space. The sequence $x_n = (x_k^n)_k$, defined by

$$x_k^n = \begin{cases} 1 & k \leq n \\ 0 & k > n \end{cases}$$

belongs to X with $\|x_n\| \leq 1$. But the pointwise limit $x = (1)_k$ does not belong to X. Thus, X is not semi-perfect.

Finally, we remark that the space given in Example 2.2.3 is a regular pre-ideal space, but not complete, i.e. not a regular ideal space.

We remark that our definition of absolute continuous norm in pre-ideal spaces differs in non-finite measure spaces from that given in [48]. But if supp$X$ is σ-finite, the set of functions of absolute continuous norm in the sense of [48] coincides with the regular part of X:

Proposition 3.3.1. *Let X be a pre-ideal space, and $x \in X$ have σ-finite support. Then x lies in the regular part of X, if and only if for any sequence $0 \leq x_n \leq |x|$, which converges a.e. monotonically decreasing to 0, we have $||x_n|| \to 0$.*

The sufficient part of this proposition is a special case of the following Lemma 3.3.2, and the necessary part is a special case of Theorem 3.3.5. In view of the latter, the proposition implies: *Over σ-finite measure spaces the regular pre-ideal spaces are precisely those pre-ideal spaces, in which Lebesgue's dominated convergence theorem holds true.*

A function x in the regular part of a pre-ideal* space satisfies $||P_D x|| \to 0$, when D tends to the empty set in some sense (for pre-ideal spaces this is also a consequence of Theorem 3.3.5):

Lemma 3.3.1. *Let X be a pre-ideal* space over some measure space $S = \bigcup S_n$ with measurable $S_1 \subseteq S_2 \subseteq \ldots$, and x be in the regular part of X. Then*

$$\lim_{n\to\infty} ||P_{S \setminus S_n} x|| = 0.$$

Observe that in Lemma 3.3.1 we did neither assume that S is σ-finite, nor that S_n has finite measure. But in σ-finite measure spaces the above condition even characterizes functions in the regular part of X, see also [48, §72 Theorem 1].

Lemma 3.3.2. *Let X be a pre-ideal* space, and $x \in X$ have σ-finite support. Then x belongs to the regular part of X, if and only if for any sequence $D_1 \supseteq D_2 \supseteq \ldots$ of measurable sets with $\bigcap D_n = \emptyset$, we have*

$$\lim_{n\to\infty} ||P_{D_n} x|| = 0.$$

Lemma 3.3.1 and 3.3.2 are special cases of Lemma 3.3.3 and 3.3.4.

Corollary 3.3.1. *Let X be a pre-ideal* space over some measure space (S, σ, μ), and x be in the regular part of X. Assume, there exists a measure ν on (S, σ) with the same null sets, such that x has σ-finite support with respect to the new measure ν. Then x belongs to the regular part of X with respect to the new measure ν.*

Proof. We have to check the condition of Lemma 3.3.2 for x. But by putting $S_n = S \setminus D_n$, Lemma 3.3.1 implies that these are satisfied. □

Corollary 3.3.2. *If (S, σ, μ) and (S, σ, ν) are both σ-finite and have the same null sets, then each regular pre-ideal* spaces over one of these measure spaces is also regular over the other.*

Corollary 3.3.2 is especially important in view of Corollary 2.2.6: It implies that we may replace the measure by a finite measure, but regular pre-ideal* spaces remain regular.

Definition 3.3.2. *$M \subseteq X$ has* uniformly absolute continuous norms, *if*

$$\lim_{\delta \to 0} \sup_{\mathrm{mes} D \leq \delta} \sup_{x \in M} \|P_D x\| = 0.$$

$M \subseteq X$ vanishes uniformly at ∞ in norm, *if*

$$\inf_{\mathrm{mes} E < \infty} \sup_{x \in M} \|P_{S \setminus E} x\| = 0.$$

A sequence $x_n \subseteq X$ has uniformly absolute continuous norms, *or* vanishes uniformly at ∞ in norm, *if the set $\{x_n : n \in \mathbb{N}\} \subseteq X$ has this property.*

The importance of these definitions will become clear in Theorem 3.3.3. But at first, we consider the definitions more precisely. Lemma 3.3.1 and 3.3.2 hold even more generally:

Lemma 3.3.3. *Let X be a pre-ideal* space, and $M \subseteq X$ have uniformly absolute continuous norms and vanish uniformly at ∞ in norm. Then for any sequence of measurable sets $D_1 \supseteq D_2 \supseteq \ldots$ with* $\mathrm{mes} \bigcap D_n = 0$ *we have*

$$\lim_{n \to \infty} \sup_{x \in M} \|P_{D_n} x\| = 0.$$

Proof. Denote the underlying measure space by S. Given $\varepsilon > 0$, there exists a set E of finite measure with

$$\|P_{S \setminus E} x\| \leq \varepsilon \qquad (x \in M).$$

Putting $E_n = E \cap D_n$, we have by $E_1 \supseteq E_2 \supseteq \ldots$ and $\mathrm{mes} E_1 < \infty$ that $\mathrm{mes} E_n \to \mathrm{mes} \bigcap E_n = 0$. Hence,

$$\|P_{E_n} x\| \leq \varepsilon \qquad (x \in M)$$

for some n. Since $|\chi_{D_n}| \leq |\chi_{S \setminus E} + \chi_{E_n}|$, we have

$$\|P_{D_n} x\| \leq \|P_{S \setminus E} x + P_{E_n} x\| \leq 2\varepsilon \qquad (x \in M)$$

for that n. □

In σ-finite measure spaces the converse is also true:

Lemma 3.3.4. *Let X be a pre-ideal* space, and $M \subseteq X$ have σ-finite support. Then M has uniformly absolute continuous norms and vanishes uniformly at ∞ in norm, if and only if for any sequence $D_1 \supseteq D_2 \supseteq \ldots$ of measurable sets with $\bigcap D_n = \emptyset$ we have*

$$\lim_{n\to\infty} \sup_{x\in M} \|P_{D_n}x\| = 0.$$

Proof. Necessity is clear by Lemma 3.3.3. Thus, assume M satisfies the condition in the theorem. Denote the measure space by S. Since $\mathrm{supp}M = \bigcup E_n$ where $E_1 \subseteq E_2 \subseteq \ldots$ have finite measure, we find, putting $D_n = \mathrm{supp}M \setminus E_n$ that

$$\inf_n \sup_{x\in M} \|P_{S\setminus E_n}x\| = \inf_n \sup_{x\in M} \|P_{D_n}x\| = 0,$$

whence M vanishes uniformly at ∞ in norm. If M has not uniformly absolute continuous norms, there exist $\varepsilon > 0$ and a sequence E_n of measurable sets with $\mathrm{mes}E_n < n^{-2}$, such that

$$\sup_{x\in M} \|P_{E_n}x\| \geq \varepsilon \qquad (n \in \mathbb{N}).$$

Putting $F_n = \bigcup_{k=n}^{\infty} E_n$, we have $\mathrm{mes}F_n \leq \sum_{k=n}^{\infty} k^{-2} \to 0$ and $F_1 \supseteq F_2 \supseteq \ldots$, whence $F = \bigcap F_n$ satisfies $\mathrm{mes}F = \lim \mathrm{mes}F_n = 0$. Putting $D_n = F_n \setminus F$, we have $D_1 \supseteq D_2 \supseteq \ldots$ and $\bigcap D_n = \emptyset$, but

$$\sup_{x\in M} \|P_{D_n}x\| = \sup_{x\in M} \|P_{F_n}x\| \geq \sup_{x\in M} \|P_{E_n}x\| \geq \varepsilon \qquad (n \in \mathbb{N}),$$

a contradiction to our assumption. □

Now we show some connections between the definitions:

Theorem 3.3.1. *Let x_n be a Cauchy sequence in some pre-ideal* space X. Then*

1. *x_n has uniformly absolute continuous norms, if and only if each x_n has absolute continuous norm.*
2. *x_n vanishes uniformly at ∞ in norm, if and only if each x_n vanishes at ∞ in norm.*

Proof. 1. Let each x_n have absolute continuous norm. Let $\varepsilon > 0$ be given. Fix n that large that $\|x_m - x_n\| < \varepsilon$ for $m \geq n$. Choose $\delta > 0$, such that $\mathrm{mes}D < \delta$ implies $\|P_D x_n\| < \varepsilon$. For all thus D and $m \geq n$ we have

$$\|P_D x_m\| = \|P_D(x_m - x_n) + P_D x_n\| \leq \|x_m - x_n\| + \|P_D x_n\| \leq 2\varepsilon.$$

Thus x_n has uniformly absolute continuous norm.
2. Let each x_n vanish at ∞ in norm. Let $\varepsilon > 0$ be given. Choose n large enough such that $\|x_m - x_n\| < \varepsilon$ for $m \geq n$. Choose a set E of finite measure with $\|P_{S\setminus E}x_n\| < \varepsilon$. Then for all $m \geq n$ we have

$$\|P_{S\setminus E}x_m\| = \|P_{S\setminus E}(x_m - x_n) + P_{S\setminus E}x_n\| \leq \|x_m - x_n\| + \|P_{S\setminus E}x_n\| \leq 2\varepsilon.$$

Thus x_n vanishes uniformly at ∞ in norm. □

All spaces X_0, X^0 and X_0^0 are ideal* spaces, if X is an ideal* space:

Theorem 3.3.2. *The inner- and outer-regular parts of any pre-ideal* space X are closed in X.*

Proof. 1. The inner-regular part X_0 is closed: Let $x_n \to x$ in X, $x_n \in X_0$. By Theorem 3.3.1, x_n has uniformly absolute continuous norm. Thus, for any $\varepsilon > 0$ there exists $\delta > 0$ such that $\mathrm{mes}D < \delta$ implies $\|P_D x_n\| < \varepsilon$ for all n. For $\mathrm{mes}D < \delta$

$$\|P_D x\| = \|P_D(x - x_n) + P_D x_n\| \leq \|x - x_n\| + \|P_D x_n\| \leq 2\varepsilon,$$

if n is large enough. Hence x has absolute continuous norm.
2. The outer-regular part X^0 is closed: Let $x_n \to x$ in X, $x_n \in X^0$. By Theorem 3.3.1, x_n vanishes uniformly at ∞ in norm. Thus for any $\varepsilon > 0$ there exists a set E of finite measure with $\|P_{S\setminus E}x_n\| < \varepsilon$ for all n. Then

$$\|P_{S\setminus E}x\| = \|P_{S\setminus E}(x - x_n) + P_{S\setminus E}x_n\| \leq \|x - x_n\| + \|P_{S\setminus E}x_n\| \leq 2\varepsilon,$$

if n is large enough. Hence x vanishes at ∞ in norm. □

It is a very important fact that a partial converse of Theorem 3.3.1 is true. We have to prepare this result:

Lemma 3.3.5. *Let $u \geq 0$ be measurable with finite support, y_n be measurable with* $\mathrm{supp}y_n \subseteq \mathrm{supp}u$, *$y_n \to 0$ in measure. Then*

$$E_n = \{s : |y_n(s)| > u(s)\}$$

satisfies $\mathrm{mes}E_n \to 0$.

Proof. Since $M_k = \{s : 0 < u(s) \leq 1/k\}$ satisfies $M_1 \supseteq M_2 \supseteq \dots$ and $\mathrm{mes}M_1 < \infty$, we have: $\mathrm{mes}M_k \to \mathrm{mes}\bigcap M_k = 0$. Thus, given any $\varepsilon > 0$, there exists $k \in \mathbb{N}$ with $\mathrm{mes}M_k < \varepsilon$. Since

$$E_n \subseteq M_k \cup \{s : |y_n(s)| \geq 1/k\},$$

we have $\mathrm{mes}E_n \leq 2\varepsilon$ for n large enough. □

Lemma 3.3.6. *Let X be an ideal space with finite support. If $x_n \in X$ converges to some x in measure and has uniformly absolute continuous norms, then $x \in X$ and $\|x - x_n\| \to 0$.*

Proof. By Corollary 2.2.4 there exists a function $u \in X, u \geq 0$ with $\text{supp}x_n \subseteq \text{supp}u$ for all n. We may assume $\|u\| = 1$. Lemma 3.3.5 implies that the measure of

$$D_n = \{s \in \text{supp}X : |x_n(s) - x(s)| > \varepsilon u(s)\}$$

tends to zero for any $\varepsilon > 0$. Thus the measure of

$$E_{nm} = \{s : |x_n(s) - x_m(s)| > 2\varepsilon u(s)\}$$

is by $E_{nm} \subseteq D_n \cup D_m$ arbitrary small, if n, m are large enough. Hence for n, m large enough we have

$$\|P_{E_{nm}} x_n\| < \varepsilon, \quad \|P_{E_{nm}} x_m\| < \varepsilon,$$

which implies by the definition of E_{nm}

$$\begin{aligned} \|x_n - x_m\| &= \|P_{E_{nm}} x_n - P_{E_{nm}} x_m + P_{S \setminus E_{nm}}(x_n - x_m)\| \\ &\leq \|P_{E_{nm}} x_n\| + \|P_{E_{nm}} x_m\| + 2\varepsilon \|u\| \leq 4\varepsilon. \end{aligned}$$

Thus x_n is Cauchy in X, and the statement follows by Corollary 2.2.3. □

Now we can prove Vitali's convergence theorem for ideal spaces (for finite measure spaces it is stated in [50, Theorem 14]):

Theorem 3.3.3. *Let X be an ideal space, and $x_n \in X$ be a sequence in the regular part of X, converging extended* to some a.e. finite function x. Then $x \in X$ and $\|x - x_n\| \to 0$, if and only if*

1. *x_n has uniformly absolute continuous norms, and*
2. *x_n vanishes uniformly at ∞ in norm.*

Proof. The necessary part follows by Theorem 3.3.1. For the sufficient part let the two conditions be satisfied. Denote the measure space by S. Given $\varepsilon > 0$, choose a set E of finite measure with $\|P_{S \setminus E} x_n\| < \varepsilon$ for all n. Let X_E be the restriction of X to functions, which vanish outside E. By Corollary 2.2.2, X_E is an ideal space, hence by Lemma 3.3.6 $P_E x_n$ converges in X_E. Thus $P_E x_n$ is Cauchy in X, i.e. for n, m large enough, we have

$$\|x_n - x_m\| = \|P_{S \setminus E} x_n - P_{S \setminus E} x_m + (P_E x_n - P_E x_m)\| \leq 3\varepsilon.$$

Thus x_n is Cauchy in X, and the statement follows by Corollary 2.2.3. □

We remark that we may not drop the condition that x is finite a.e.:

Example 3.3.1. Let $S = \{0\}$, $\text{mes}S = 1$, and X consist of all real functions, $\|x\| = |x(0)|$. Then $x_n(0) = n$ has uniformly absolute continuous norms, but the extended limit is not finite (and not in X).

However, if x_n is a priori bounded in norm, or if we work in the Lebesgue measure, we do not need to know a priori that x is finite:

Call a measurable set E *finite atomic free*, if any subset of E of finite measure can be divided into two parts of equal measure.

Lemma 3.3.7. *Let X be a pre-ideal space. Let $x_n \in X$ have uniformly absolute continuous norms, such that each $E_n = \operatorname{supp} x_n$ is finite atomic free. Then, if $\operatorname{mes} E_n$ is bounded, x_n is bounded in norm.*

Proof. Let $S = \sup \operatorname{mes} E_n$. Choose N such that $\operatorname{mes} D \le 2^{-N} S = \delta$ implies $\|P_D x_n\| \le 1$ for all n. Now fix some n. Since E_n is finite atomic free, there exist pairwise disjoint sets $D_1^n, \ldots, D_{2^N}^n$ with $E_n = \bigcup_j D_j^n$ and $\operatorname{mes} D_j^n = 2^{-N} \operatorname{mes} E_n \le \delta$. Then

$$\|x_n\| = \|\sum_{j=1}^{2^N} P_{D_j^n} x_n\| \le \sum_{j=1}^{2^N} \|P_{D_j^n} x_n\| \le 2^N.$$

Since n was arbitrary, this implies the statement. □

Theorem 3.3.4. *Let X be an ideal space, $x_n \in X$, and $x_n \to x$ extended*. Assume, x_n has uniformly absolute continuous norms and vanishes uniformly at ∞ in norm. Assume, moreover, that either x_n is bounded in norm, or that each $\operatorname{supp} x_n$ is finite atomic free. Then x is finite a.e., $x \in X$, and $\|x - x_n\| \to 0$.*

Proof. Let E, X_E be as in the proof of Theorem 3.3.3. If we can show that $P_E x_n$ is a Cauchy-sequence in X_E, we may continue as in the proof there. By Lemma 3.3.6 it suffices to prove that x is finite a.e. on E. If this is false, Lemma 2.2.3 implies that there exists a subsequence $y_k = P_E x_{n_k}$ with $|y_k(s)| \to \infty$ for all s in some subset $F \subseteq E$ of positive measure. But on the other hand, by Lemma 3.3.7, y_k is bounded in norm. Thus we have a contradiction by Lemma 2.2.2. □

Observe that Theorems 3.3.3 and 3.3.4 both fail, if you drop in the last case of Definition 2.2.3 the condition that $\operatorname{supp} x$ and each $\operatorname{supp} x_n$ have the finite subset property, even if the measure space is the union of sets of finite measure. You can use the same counterexamples as in Example 2.2.1.

A consequence of Vitali's theorem is of course Lebesgue's dominated convergence theorem, since $|x_n| \le y$ implies $\|P_D x_n\| \le \|P_D y\|$ for all n and all measurable D. However, that theorem is even true for pre-ideal spaces, and holds also, if $x_n \to x$ in measure on each set of finite measure, even if $\operatorname{supp} x$ or some $\operatorname{supp} x_n$ does not have the finite subset property. For σ-finite measure spaces the theorem is proved by different means in [48, §72 Theorem 1/2]:

Theorem 3.3.5. *Let x_n be a sequence of measurable functions, such that $x_n \to x$ a.e. or in measure on each set of finite measure. Assume, $|x_n| \leq |y|$ and $|x| \leq |y|$, where y lies in the regular part of a pre-ideal space X. Then $x_n, x \in X$ and $||x - x_n|| \to 0$.*

Proof. Denote the measure space by S. Given $\varepsilon > 0$, there exists a set E of finite measure, such that

$$||P_{S\setminus E}y|| < \varepsilon. \tag{3.8}$$

Define

$$E_n = \{s \in E : |x(s) - x_n(s)| > \varepsilon\,|y(s)|\}.$$

Since $x_n \to x$ in measure on E, Lemma 3.3.5 implies mes$E_n \to 0$. Since y has absolute continuous norm, we have

$$||P_{E_n}y|| < \varepsilon,$$

if n is large enough. Together with (3.8) and the definition of E_n this implies for all those n:

$$\begin{aligned} ||x - x_n|| &= ||P_{S\setminus E}(x - x_n) + P_{E_n}(x - x_n) + P_{E\setminus E_n}(x - x_n)|| \\ &\leq 2||P_{S\setminus E}y|| + 2||P_{E_n}y|| + ||\varepsilon\,|y|\;|| \leq (4 + ||y||)\varepsilon, \end{aligned}$$

thus $||x - x_n|| \to 0$. □

The assumption $|x| \leq |y|$ is not necessary, if $x_n \to x$ extended*, since in this case it follows by $|x_n| \leq |y|$ and Lemma 2.2.4. But in general we may not drop this assumption, even if the measure space is the union of sets of finite measure, as can be seen by $x_n = 0$, $x = 1$ in Example 2.2.1.

A consequence of Theorem 3.3.5 is Beppo Levi's monotone convergence theorem for ideal spaces:

Corollary 3.3.3. *Let X be a completely regular real-valued ideal space, $x_n \in X$ be bounded. If $x_1 \leq x_2 \leq \ldots$ or $x_1 \geq x_2 \geq \ldots$ a.e., then x_n converges a.e. to some $x \in X$, and $||x - x_n|| \to 0$.*

Proof. $y_n = x_n - x_1 \in X$ converges a.e. in the extended sense to some y, and has bounded norm. Since X is perfect, we have $y \in X$, hence $x_n \to x = y - x_1 \in X$. Since $|y_n| \leq |y|$, Theorem 3.3.5 now implies $||x - x_n|| = ||y - y_n|| \to 0$. □

For Corollary 3.3.3 it is not enough that X is just a regular ideal space, as can be seen by the space $X = c_0$ of sequences converging to zero, equipped with the sup-norm: The sequence $x_1 = (1, 0, 0, \ldots)$, $x_2 = (1, 1, 0, 0, \ldots)$, $x_3 = (1, 1, 1, 0, 0, \ldots)$, $\ldots$ is bounded and converges monotonically increasing a.e. to $x = (1, 1, \ldots)$, but $x \notin X$. In other words: You can not prove the monotone convergence theorem by Vitali's theorem and Lebesgue's dominated convergence theorem without using special properties of X. This shows that the proof of [15, Corollary III.6.17] needs some completion (indeed, the proof uses the fact that the estimate

$$\int_S f_n(s)\mu(ds) \leq \int_S f_N(s)\mu(ds) + \varepsilon$$

implies

$$\limsup_{\mu(E)\to 0} \int_E f_n(s)\mu(ds) \leq \limsup_{\mu(E)\to 0} \int_E f_N(s)\mu(ds) + \varepsilon \text{ uniformly in } n,$$

which is not true in general without other assumptions).

The following corollary is proved in [50, Theorem 18] for ideal spaces only:

Corollary 3.3.4. *Each regular pre-ideal space is almost perfect.*

Proof. We use Lemma 3.2.7. Let $0 \leq x_n \in X$ be monotonically increasing, and $x = \sup x_n \in X$. By Theorem 3.3.5 we have $\|x - x_n\| \to 0$, whence $\|x_n\| \to \|x\|$. □

We establish a theorem, which will allow us to estimate $\|x - x_n\|$ by just estimating $\|x_n\|$ in regular spaces, in which the norm is 'almost additive' for complemented projections:

Theorem 3.3.6. *Let X be a pre-ideal space over some measure space S, and $x_n \in X$ converging in measure on each set of finite measure to some function x, which lies in the regular part of X. Let $c(u, v)$ be continuous on $0 \leq v \leq u \leq \sup \|x_n\| < \infty$, nondecreasing in u and and nonincreasing in v, such that*

$$\|P_{S\setminus E}x_n\| \leq c(\|x_n\|, \|P_E x_n\|) \qquad (\text{mes}E < \infty,\ n = 1, 2, \ldots). \tag{3.9}$$

Then we have

$$L = \limsup_{n\to\infty} \|x_n\| \geq \|x\|, \tag{3.10}$$

and

$$\limsup_{n\to\infty} \|x - x_n\| \leq c(L, \|x\|). \tag{3.11}$$

Proof. Without loss of generality, assume $\|x - x_n\| \to \limsup \|x - x_n\|$ (otherwise consider a subsequence with that property, and observe that it suffices to prove the statement for that subsequence).
Given $\varepsilon > 0$, choose a set $F \subseteq \text{supp}X$ of finite measure with $\|P_{S\setminus F}x\| < \varepsilon$, and $\delta > 0$, such that $\text{mes}D \leq \delta$ implies $\|P_D x\| < \varepsilon$. Since $x_n \to x$ on F in measure, we have $x_{n_k} \to x$ a.e. on F for some proper subsequence. By

Egorov's theorem there exists a measurable $H \subseteq F$ with $\operatorname{mes}(F \setminus H) < \delta$, such that $x_{n_k} \to x$ uniformly on H. Furthermore, by Theorem 2.2.5, there exists some $E \subseteq H$ with $\operatorname{mes}(H \setminus E) < \delta$, such that χ_E belongs to the real form of X. We have

$$\|P_{S\setminus E}x\| = \|P_{S\setminus F}x + P_{F\setminus H}x + P_{H\setminus E}x\| \leq 3\varepsilon. \tag{3.12}$$

Since $x_{n_k} \to x$ uniformly on E, we have $s_k = \operatorname{ess\,sup} |P_E x(s) - P_E x_{n_k}(s)| \to 0$. Hence, for k large enough,

$$\|P_E(x - x_{n_k})\| \leq \|s_k \chi_E\| \leq \varepsilon. \tag{3.13}$$

By (3.12) and (3.13) we have

$$\|x\| = \|P_{S\setminus E}x + P_E(x - x_{n_k}) + P_E x_{n_k}\| \leq 4\varepsilon + \|P_E x_{n_k}\|. \tag{3.14}$$

By $\|x_{n_k}\| \geq \|P_E x_{n_k}\|$ this implies (3.10). Now combine the previous formulas and (3.9) to see that

$$\begin{aligned} \|x - x_{n_k}\| &= \|P_E(x - x_{n_k}) + P_{S\setminus E}x - P_{S\setminus E}x_{n_k}\| \leq 4\varepsilon + \|P_{S\setminus E}x_{n_k}\| \\ &\leq 4\varepsilon + c(\|x_{n_k}\|, \|P_E x_{n_k}\|) \leq 4\varepsilon + c(\|x_{n_k}\|, \max\{\|x\| - 4\varepsilon, 0\}), \end{aligned}$$

if k is large enough. This implies the statement, since $c(u,v)$ is uniformly continuous for $0 \leq v \leq u \leq \sup \|x_{n_k}\|$ and since $\|x - x_{n_k}\| \to \limsup \|x - x_n\|$. □

You may not drop the condition that x lies in the regular part of X. In fact, in $L_\infty([0,1])$ the following easy corollary fails for $x_n(s) = 2s^n$, $x \equiv 0$, $y \equiv 1$:

Corollary 3.3.5. *Let X be regular, $x_n \in X$ converging in measure on each set of finite measure to some $x \in X$. Then*

$$L(y) = \limsup_{n\to\infty} \|y - x_n\|$$

attains its smallest value for $y = x$.

Proof. Since $z_n = x_n - y \to z = x - y$, Theorem 3.3.6 implies for $c(u,v) = u$ that $L(x) = \limsup \|z - z_n\| \leq \limsup \|z_n\| = L(y)$. □

However, in many spaces you may use smaller functions than $c(u,v) = u$. As a sample application we state a result similar to [48, §30 Exercise 21] (parts of the proof of Theorem 3.3.6 can also be found there):

Corollary 3.3.6. *Let $X = L_p(S,Y)$, $1 \leq p < \infty$. Let $x_n \in X$ converge in measure on each set of finite measure to $x \in X$. Then*

$$\limsup_{n\to\infty} \|x - x_n\|^p \leq \limsup_{n\to\infty} \|x_n\|^p - \|x\|^p.$$

In particular, $\|x - x_n\| \to 0$, if and only if $\limsup \|x_n\| \leq \|x\|$.

Proof. You may choose $c(u,v) = (u^p - v^p)^{1/p}$, since for all $x \in X$ we have $\|P_E x\|^p + \|P_{S \setminus E} x\|^p = \|x\|^p$. □

From Corollary 3.3.6 you might guess that for any regular ideal space X we have that $x_n \to x$ a.e. with $x_n, x \in X$ and $\limsup \|x_n\| \leq \|x\|$ implies $\|x - x_n\| \to 0$. This is not true in general:

Example 3.3.2. Let $X = c_0$, $x = (a_k)_k \in X$, with $x \neq 0$. Define a sequence $x_n \in X$ by $x_n = (b_k^n)_k$, where

$$b_k^n = \begin{cases} a_k & \text{if } k < n, \\ \|x\| & \text{if } k = n, \\ 0 & \text{if } k > n. \end{cases}$$

Then $x_n, x \in X$, $x_n \to x$ a.e., and $\|x_n\| \leq \|x\|$. However, $\|x_n - x\| \geq |a_n - \|x\|\,| \to \|x\| \neq 0$. It is illuminating to check elementary that Corollary 3.3.5 holds anyway.

3.4 Associate Spaces

Definition 3.4.1. *If X is a pre-ideal space over some measure space S, then its* associate space X' *consists of all measurable functions, vanishing outside* $\underline{\text{supp}}X$, *for which the norm*

$$\|y\|_{X'} = \sup_{\|x\|_X \leq 1} \int_S |y(s)|\,|x(s)|\,ds \tag{3.15}$$

is finite.

We agree to use the abbreviation (3.15) also, if y is just measurable and does not necessarily vanish outside $\underline{\text{supp}}X$. But our definition ensures that (3.15) defines a norm on X'. Moreover, X' becomes perfect (see also [48, §68 Theorem 1]):

Theorem 3.4.1. *If X is a pre-ideal space, then X' is a perfect ideal space.*

Proof. At first, we prove that $\|y\|_{X'} = 0$ implies $y = 0$ a.e.: Let D be some representation of supp$y$. Since y vanishes a.e. outside $\underline{\text{supp}}X$ we have that $D \setminus \underline{\text{supp}}X$ is contained in some set N of measure zero. Hence, $E = D \setminus N$ is another representation of supp$y$, but $E \subseteq \underline{\text{supp}}X$. If $y \neq 0$, E has positive measure, and thus by Lemma 2.2.8, there exists some $x \in X$, such that $E \cap \text{supp}x$ has positive measure, which is not possible by $\|y\|_{X'} = 0$. Thus X' is a pre-ideal space.
By Corollary 3.2.4 it now suffices to show the perfectness of X'.
Let $y_n \in X'$, $\|y_n\|_{X'} \leq C < \infty$, and $y_n \to y$ a.e. in the extended sense. Then for any $x \in X$, $\|x\|_X \leq 1$, Fatou's lemma implies

$$\int_S |y(s)|\,|x(s)|\,ds \leq \liminf_{n\to\infty} \int_S |y_n(s)|\,|x(s)|\,ds \leq \liminf_{n\to\infty} \|y_n\|_{X'}\,.$$

Thus $\|y\|_{X'} \leq \liminf \|y_n\|_{X'}$. $y \in X'$ now follows by supp$y \subseteq \bigcup \text{supp}y_n$. □

In ideal spaces we may simplify the definition:

Theorem 3.4.2. *If X is an ideal space over S, then a measurable function y satisfies $\|y\|_{X'} < \infty$, if and only if*

$$\int_S |y(s)|\,|x(s)|\,ds < \infty \qquad (x \in X).$$

Proof. Without loss of generality, let X be real-valued. Fix some y with the described property and define a linear functional l on X by

$$lx = \int_S |y(s)|\,x(s)ds \qquad (x \in X).$$

We have to prove that l is continuous at 0. It suffices to prove that any sequence $x_n \to 0$ contains a subsequence with $lx_{n_k} \to 0$. Choose the subsequence such that $x_{n_k} \to 0$ a.e. and $|x_{n_k}| \le |z|$ for some $z \in X$ (Corollary 3.2.1). Since lz is defined, $w = |y|\, z$ is integrable. Since $|y|\, |x_{n_k}| \le w$, Lebesgue's dominated convergence theorem implies $lx_{n_k} \to 0$. □

Corollary 3.4.1. *If X and Y are ideal spaces over S and coincide as sets, then X' and Y' coincide as sets.*

If X is Y-valued, and X' is Y^*-valued, X' is continuously embedded in the dual space X^* of X:

Definition 3.4.2. *Let X be a Y-valued pre-ideal space, and X' be Y^*-valued. Then $y^* \in X^*$ is called* integral functional, *if there is some $y \in X'$ with*

$$y^*(x) = \int_S y(s)x(s)ds \qquad (x \in X). \tag{3.16}$$

For any $y \in X'$ we have $y^* \in X^*$: Indeed, $s \mapsto y(s)x(s)$ is measurable by Theorem A.1.2, and $\|y^*\| \le \|y\|_{X'}$. We will soon see that even equality holds (Theorem 3.4.4).

But first, we prove a theorem analogous to 3.4.2 for this situation (see also Corollary 3.4.2). The following proof uses ideas of [48, §71 Theorem 5], which gives a different method to prove Theorem 3.4.2:

Theorem 3.4.3. *Let X be a Y-valued pre-ideal* space over some measure space S with the Riesz-Fischer property. If for some measurable $y : S \to Y^*$ we have*

$$\int_S y(s)x(s)ds < \infty \qquad (x \in X,\ yx \ge 0),$$

then (3.16) *defines a bounded linear functional on X.*

Proof. First, observe that yx is measurable by Theorem A.1.2 for any $x \in X$. If the conclusion is false, there exists a sequence $z_n \in X$, $\|z_n\| \le 1$ with

$$\int_S |y(s)z_n(s)|\, ds \ge n^3.$$

Now define $x_n(s) = z_n(s)\mathrm{sgn}[\overline{y(s)z_n(s)}]$, where for complex $z \ne 0$ we define $\mathrm{sgn} z = z/\,|z|$. Then $x_n \in X$, $\|x_n\| \le 1$, $yx_n = |yz_n| \ge 0$, and

$$\int_S y(s)x_n(s)ds \ge n^3.$$

Since $\sum k^{-2}\, \|x_k\| < \infty$, the Riesz-Fischer property implies that $\sum k^{-2}x_k$ converges extended* to some $x \in X$. By Lemma 2.2.4 for almost all s there is a subsequence of the partial sums $w_n = \sum_{k=1}^{n} k^{-2}x_k$ with $w_{n_m}(s) \to x(s)$, i.e.

$$y(s)x(s) = \lim_{m\to\infty} \sum_{k=1}^{n_m} k^{-2}y(s)x_k(s).$$

By $y(s)x_k(s) \geq 0$ this implies that we even have

$$y(s)x(s) = \sum_{k=1}^{\infty} k^{-2}y(s)x_k(s)$$

for almost all s. By $yx_k \geq 0$ we have $yx \geq 0$ and

$$\int_S y(s)x(s)ds \geq \int_S n^{-2}y(s)x_n(s)ds \geq n$$

for any n, which is a contradiction to our assumption. □

We emphasize that Theorem 3.4.3 implies in view of Theorem 3.2.1 that the functional (3.16) is bounded (i.e. continuous), whenever its domain of definition is an ideal space (i.e. for ideal spaces the conditions $y^* \in X^*$, $y \in X'$ in Definition 3.4.2 are satisfied automatically).
More results in this direction will be given in Sect. 5.1.

For $X = L_p(S, Y)$ $(1 < p \leq \infty)$ parts of the following consequence of Corollary A.2.3 are stated in [48, §31 Exercise 2]. In this case the proof simplifies, since X' is regular, whence the simple functions are dense in it.

Theorem 3.4.4. *Let X be a Y-valued pre-ideal space over some measure space S. Then any measurable function $y : S \to Y^*$ satisfies*

$$\|y\|_{X'} = \sup_{\|x\|=1,\ yx\geq 0} \int_S y(s)x(s)ds,$$

where the case that a side is infinite is not excluded.
Furthermore, $\|y\|_{X'} < \infty$, if and only if (3.16) defines a bounded linear functional on X, and in this case $\|y\|_{X'} = \|y^*\|$.

Proof. For the first statement, use Corollary A.2.3 for $M = \{x \in X : \|x\| = 1\}$. Furthermore, if (3.16) is defined and bounded on X, we have

$$\sup_{x\in M,\ yx\geq 0} \int_S y(s)x(s)ds = \sup_{x\in M,\ yx\geq 0} y^*(x) \leq \sup_{x\in M} |y^*(x)|,$$

which implies $\|y\|_{X'} \leq \|y^*\| < \infty$. Conversely, if $\|y\|_{X'} < \infty$, we have that $y^* \in X^*$ with $\|y^*\| \leq \|y\|_{X'}$ by Theorem A.1.2. □

An analogous result for finite measure spaces can be found in [51, Lemma 3]. Now we can sharpen Theorem 3.4.2:

Corollary 3.4.2. *Let X be a Y-valued ideal space over some measure space S. Then for a measurable $y : S \to Y^*$ we have $\|y\|_{X'} < \infty$, if and only if*

$$\int_S y(s)x(s)ds < \infty \qquad (x \in X,\ yx \geq 0).$$

Proof. Theorem 3.4.3 implies that (3.16) defines a bounded linear functional on X. Now use Theorem 3.4.4. □

Usually, X' is not trivial, and $\mathrm{supp}X'$ is the union of all subsets of $\underline{\mathrm{supp}}X$ of finite measure. We will prepare the corresponding result:

Lemma 3.4.1. *Let H be a pre-Hilbert space with inner product $(\cdot,\cdot)$, $N \subseteq H$ be convex, $x_0 \in H \setminus N$, and $y_0 \in N$ with $\mathrm{dist}(x_0, N) = \|x_0 - y_0\|$. Then*

$$\mathrm{Re}(x, x_0 - y_0) \leq \mathrm{Re}(y_0, x_0 - y_0) \qquad (x \in N).$$

Proof. For $x \in N$ and $0 < \lambda \leq 1$ we have $z = (1-\lambda)y_0 + \lambda x \in N$. Hence

$$\begin{aligned} \mathrm{dist}(x_0, N)^2 &\leq \|x - z\|^2 = \|x_0 - y_0 - \lambda(x - y_0)\|^2 \\ &= \|x_0 - y_0\|^2 - 2\lambda \mathrm{Re}(x - y_0, x_0 - y_0) + \lambda^2 \|x - y_0\|^2 \end{aligned}$$

implies $2\mathrm{Re}(x - y_0, x_0 - y_0) \leq \lambda \|x - y_0\|^2$, which yields $\mathrm{Re}(x - y_0, x_0 - y_0) \leq 0$ for $\lambda \to 0^+$. □

Corollary 3.4.3. *Let H be a Hilbert space with inner product $(\cdot,\cdot)$, $N \subseteq H$ be closed and convex, and $x_0 \in H \setminus N$. Then there exists some $z_0 \in H$, such that*

$$\sup_{x \in N} \mathrm{Re}(x, z_0) < \mathrm{Re}(x_0, z_0).$$

Proof. Choose y_0 as in Lemma 3.4.1, and define $z_0 = x_0 - y_0$. Then for $x \in N$ we have $\mathrm{Re}(x, z_0) \leq \mathrm{Re}(y_0, z_0) < \mathrm{Re}(x_0, z_0)$, where the last inequality follows by $\mathrm{Re}(x_0, z_0) - \mathrm{Re}(y_0, z_0) = \|z_0\|^2 > 0$. □

Of course, Corollary 3.4.3 might just be considered as an application of Riesz's representation theorem to the separation theorem. But we did not want to refer to the latter, since its usual proof uses Hahn-Banach's theorem (see Remark 2.1.1).

Now we are able to prove the following fundamental result, which mainly is due to Lorentz (see [48, §71 Theorem 2] and [51, Theorem 6]; cf. also [50, Lemma 4] and [50, Theorem 21]). We write

$$\|x\|_{X''} = \sup_{\|y\|_{X'} \le 1} \int_S |x(s)|\,|y(s)|\,ds.$$

Recall that $\|y\|_{X'}$ is defined, even if y does not vanish a.e. outside $\underline{\text{supp}}X$.

Theorem 3.4.5. *Let X be a pre-ideal space, and $x \neq 0$ be measurable, where* supp$x$ *is a σ-finite subset of* $\underline{\text{supp}}X$. *Then*

$$\|x\|_{X''} = \|x\|_L > 0,$$

where $\|x\|_L$ denotes the Lorentz seminorm of x.

Proof. Without loss of generality, let X, X' be real-valued, and $x \ge 0$. Denote the measure space by S. At first, we prove $\|x\|_{X''} \le \|x\|_L$:
Let $x_n \in X$ converge a.e. in the extended sense to x. We have to prove that $\|x\|_{X''} \le \liminf \|x_n\| = L$. For any fixed y satisfying $\|y\|_{X'} \le 1$ Fatou's lemma implies

$$\int_S |x(s)|\,|y(s)|\,ds \le \liminf_{n\to\infty} \int_S |x_n(s)|\,|y(s)|\,ds \le \liminf_{n\to\infty} \|x_n\| = L,$$

whence $\|x\|_{X''} \le L$ as stated.
To prove $\|x\|_L \le \|x\|_{X''}$, first assume additionally that $E = \text{supp}x$ has finite measure, that x is bounded, and that $x \in X$. By multiplying with a proper constant we may assume $\|x\| = 1$. Let $L_2 = L_2(E)$,

$$M = \{z \in L_2 : z = y|_E \text{ for some } y \in X \text{ with } \|y\| \le 1\},$$

and $N = \overline{M}$ be the closure of M in L_2. Corollary 3.2.7 implies $L = \|x\|_L > 0$. Thus for $\varepsilon > 0$ we may define $x_0 = (L^{-1} + \varepsilon)x$. We state that $x_0|_E \notin N$: Indeed, otherwise there exists a sequence $x_n \in M$ with $\|x_n - x_0|_E\|_{L_2} \to 0$. Since this implies that the trivial extension of x_n to the whole measure space converges extended to x_0, we have by Remark 3.2.2 the contradiction $\|x_0\|_L \le 1$.
Thus $x_0|_E \notin N$. Applying Corollary 3.4.3 to $H = L_2$ we find some $z_0 \in H$ satisfying

$$\sup_{z\in N}(z, z_0) < (x_0|_E, z_0) = (L^{-1} + \varepsilon)(x|_E, z_0) = c. \tag{3.17}$$

Let w be the trivial extension of z_0 to the whole measure space. By $0 \in N$ we have $c > 0$, i.e. $\|w\|_{X'} \ge (x|_E, z_0) > 0$. We now claim $\|w\|_{X'} \le c$. Indeed, given $z \in X$, $\|z\| \le 1$, define $z_n(s) = \min\{|z(s)|, n\}\text{sgn}w(s)$. By $z_n|_E \in M$ and (3.17) we have

$$\int_S |w(s)|\,|z_n(s)|\,ds = (z_n|_E, z_0) < c.$$

Since $|w|\,|z_n| \to |w|\,|z|$ a.e., Fatou's lemma implies

$$\int_S |w(s)|\,|z(s)|\,ds \le c,$$

hence $\|w\|_{X'} \le c$ as stated. Now set $y = w/\|w\|_{X'}$. Then $\|y\|_{X'} = 1$ and

$$\int_S |x(s)|\,|y(s)|\,ds \ge c^{-1}\int_S |x(s)|\,|w(s)|\,ds \ge c^{-1}(x|_E, z_0) = (L^{-1}+\varepsilon)^{-1},$$

hence $\|x\|_{X''} \ge (L^{-1}+\varepsilon)^{-1}$, which implies the statement.
Now, we drop the additional assumptions. Let X_E be the restriction of X to functions vanishing outside $E = \text{supp}x$. Lemma 2.2.8 implies $\text{supp}X_E = E$. Hence, by Corollary 2.2.7 there exists a sequence of sets $E_1 \subseteq E_2 \subseteq \ldots$ of finite measure with $\bigcup E_n = E$ and $\chi_{E_n} \in X$. Now consider the sequence $x_n = \min(x, n\chi_{E_n})$. Then x_n satisfies $\|x_n\|_{X''} = \|x_n\|_L$ as we have seen, and converges a.e. monotonically increasing to x in the extended sense. In view of Lemma 3.2.9 this implies $\|x\|_L \le \liminf \|x_n\|_L = \liminf \|x_n\|_{X''} \le \|x\|_{X''}$. □

The most important consequence of Theorem 3.4.5 is, as we mentioned before, that $\text{supp}X' = \text{supp}X$, if $\text{supp}X$ is σ-finite. See also [50, Lemma 4] (we remark that this apparently more general lemma is in view of Lemma 3.2.10 in fact a special case of Theorem 3.4.6):

Theorem 3.4.6. *Let X be a pre-ideal space. Then for any subset $E \subseteq \underline{\text{supp}}X$ of σ-finite measure there exists some $y \in X'$ with $E = \text{supp}y$.*

Proof. By Theorem 3.4.1 and Theorem 2.2.2 the restriction X'_E of X' to functions vanishing outside E is an ideal space. Hence, by Theorem 2.2.6 it suffices to prove that for any subset $D \subseteq E$ of positive measure there exists some function $y \in X'$ with $\text{mes}(D \cap \text{supp}y) > 0$. To see this, choose a nontrivial $x \in X$ with $\text{supp}x \subseteq D$. Then Theorem 3.4.5 implies $\|x\|_{X''} > 0$, i.e. there is some z satisfying $\|z\|_{X'} \le 1$ and $\int_D |x(s)|\,|z(s)|\,ds > 0$. Now just put $y = P_E z$. □

The theorem is complemented by:

Theorem 3.4.7. *Let X be a pre-ideal space. Let $M = \overline{\text{supp}}X$, and U be the union of all subsets $C \subseteq M$ with $\text{mes}C < \infty$. Then $\overline{\text{supp}}X' \cap M \subseteq U$.*

Proof. Assume the contrary, i.e. $E = (\overline{\text{supp}}X' \cap M) \setminus U$ contains some point s_0. Since $s_0 \in \overline{\text{supp}}X'$ there is some $y \in X'$ such that $s_0 \in D$ for a proper representation D of $\text{supp}y$, i.e. we may assume $y(s_0) \ne 0$. Also there exists a representation of some $x \in X$ with $x(s_0) \ne 0$ and $\text{supp}x \subseteq M$.
Now observe that by $s_0 \notin U$ we have that no $C \subseteq M$ with $s_0 \in C$ may satisfy $\text{mes}C < \infty$. Since the set

$$C = \{s : |x(s)|\,|y(s)| \ge |x(s_0)|\,|y(s_0)|\} \subseteq \text{supp}x \subseteq M$$

belongs to the Lebesgue-extension of the measure with $s_0 \in C$, we thus must have mes$C = \infty$, whence (S denoting the measure space)

$$\int_S |y(s)|\,|x(s)|\,ds \geq \int_C |x(s_0)|\,|y(s_0)|\,ds = \infty,$$

a contradiction to $x \in X$ and $y \in X'$. □

Observe that even if supp$X$ and suppX' both exist, and the union U of all subsets of suppX' of finite measure is measurable, we need not have that supp$X' = U$ (this shows that Theorem 3.4.6 does not imply even in this special case that $\underline{\text{supp}}X'$ or $\overline{\text{supp}}X'$ must contain U):

Example 3.4.1. Let S be some uncountable set. Let each $E \subseteq S$ be measurable. If E is at most countable let mes$E = 0$, otherwise mes$E = \infty$. Let $X = L_\infty(S)$. Then $X' = \{0\}$, hence supp$X' = \emptyset$, but the union of all sets $E \subseteq$ supp$X$ of finite measure is S.

The following consequences generalize [48, §71 Theorem 3(a)/(c)] and parts of [50, Theorem 22]:

Theorem 3.4.6 states that X' usually is not trivial. But Theorem 3.4.5 implies even more, namely that usually X' is large enough to characterize X:

Corollary 3.4.4. *Let X be an almost α-perfect pre-ideal space, and $x \in X$ have σ-finite support. Then*

$$\alpha^{-1}\,\|x\| \leq \|x\|_{X''} \leq \|x\|\,.$$

Proof. The second inequality is trivial. For the first inequality, assume without loss of generality that supp$X$ = supp$x$ (otherwise consider the restriction of X to functions vanishing outside supp$x$). Since X is almost α-perfect, we have $\|x\| \leq \alpha\,\|x\|_L$. Now apply Theorem 3.4.5. □

Corollary 3.4.5. *Let X be a semi-perfect pre-ideal space. Let x be measurable, and* supp$x$ *be a σ-finite subset of* $\underline{\text{supp}}X$. *Then $\|x\|_{X''} < \infty$ implies $x \in X$.*

Proof. Since X is semi-perfect, $\|x\|_L < \infty$ implies $x \in X$. Thus the statement follows by Theorem 3.4.5. □

The following theorems show that the previous corollaries may not be sharpened. We define $X'' = (X')'$. If supp$X$ is σ-finite, then suppX' = supp$X$ by Theorem 3.4.6. Thus by Theorem 3.4.5 we have $X'' = X_L$ (with the same norms). In particular, $X \subseteq X''$ and X'' is the 'perfect hull' of X.

Theorem 3.4.8. *Let X be a pre-ideal space with σ-finite support. Then we have*

$$\alpha^{-1} \|x\| \leq \|x\|_{X''} \leq \|x\| \qquad (x \in X)$$

if and only if X is almost α-perfect.

Proof. Sufficiency follows by Corollary 3.4.4. For necessity, assume X satisfies $\|x\| \leq \alpha \|x\|_{X''}$ for some $\alpha \geq 1$. Let $x_n \in X$ converge a.e. to some $x \in X$. By Theorem 3.4.5 we have $\|x\| \leq \alpha \|x\|_L \leq \alpha \liminf \|x_n\|$, i.e. X is almost α-perfect. □

Theorem 3.4.9. *Let X be a pre-ideal space with σ-finite support. Then we have $X'' \subseteq X$, if and only if X is semi-perfect. Moreover, in this case $X = X''$ with equivalent norms.*

Proof. We first prove necessity. Thus, assume $X'' \subseteq X$ and let $x_n \in X$ converge a.e. to x in the extended sense with $\|x_n\| \leq c$. Then $\|x\|_L \leq c$, whence $\|x\|_{X''} \leq c$ by Theorem 3.4.5. Since by Theorem 3.4.6 we have $\operatorname{supp} x \subseteq \operatorname{supp} X = \operatorname{supp} X'$, this implies $x \in X''$, whence $x \in X$ by assumption.
Sufficiency follows by Corollaries 3.4.4, 3.4.5, and Theorem 3.2.2. □

Corollary 3.4.6. *Let X be a pre-ideal space with σ-finite support. Then $X = X''$ with the same norms, if and only if X is perfect.*

Example 3.4.1 shows that in all previous results we may not drop the condition of $\operatorname{supp} x$ or $\operatorname{supp} X$ being σ-finite.

We emphasize the following consequence of Corollary 3.4.4 and Corollary A.2.4:

Corollary 3.4.7. *Let Y be a Banach space with the bidual property, and X be a Y-valued almost perfect pre-ideal space over some measure space S. Then for any x with σ-finite support and any $\varepsilon > 0$ there exists some $y^* \in X^*$, $\|y^*\|_{X^*} \leq 1 + \varepsilon$ with $y^*(x) = \|x\|$.*
Moreover, y^* can be chosen to be of the form (3.16), *where $y : S \to Y^*$ is measurable, $yx \geq 0$, and $\|y\|_{X'} \leq 1 + \varepsilon$.*

Proof. Let M consist of all measurable functions $z : S \to Y^*$ with $\|z\|_{X'} \leq 1$. Corollary 3.4.4 and Corollary A.2.4 then imply

$$\|x\|_X = \|x\|_{X''} = \sup_{z \in M,\ zx \geq 0} \int_S z(s)x(s)ds.$$

Hence, there exists some $z \in M$ with $zx \geq 0$ and

$$(1 + \varepsilon)^{-1} \|x\|_X \leq \int_S z(s)x(s)ds = \alpha.$$

Hence, for $\alpha = 0$ choose $y = 0$, otherwise $y(s) = \alpha^{-1} \|x\|_X z(s)$. □

Corollary 3.4.8. *Let Y be a Banach space with the bidual property, and X be a Y-valued almost perfect pre-ideal space.*

1. *If the support of some function in X contains a set of finite positive measure, then the ε-duality map is nontrivial for any $\varepsilon > 0$.*
2. *If $\operatorname{supp} X$ is σ-finite, then X has the bidual property.*

The following important consequence of Theorem 3.4.6 will need a lemma, which is well-known in integration theory (see e.g. [15, Lemma III.6.8]):

Lemma 3.4.2. *Let x be integrable with values in $[0, \infty]$ or in some Banach space Y. If*

$$\int_E x(s)ds = 0$$

for all measurable sets E, then $x = 0$ a.e.

Theorem 3.4.10. *Let X be a pre-ideal space over some measure space S. If $x : S \to [0, \infty]$ or $x : S \to Y$ (Y being some Banach space) is measurable with*

$$\int_S y(s)x(s)ds = 0 \qquad (y \in X',\ y \geq 0),$$

then x vanishes a.e. on every subset of $\underline{\operatorname{supp}} X$ of σ-finite measure.

Proof. Let $E \subseteq \underline{\operatorname{supp}} X$ have σ-finite measure. By Theorem 3.4.6 there is some $y \in X'$, $y \geq 0$ with $\operatorname{supp} y = E$. Hence $z(s) = y(s)x(s)$ is integrable, and for all measurable sets D our assumption implies for $y_D = P_D y \in X'$ that

$$\int_D z(s)ds = \int_S y_D(s)x(s)ds = 0.$$

Thus Lemma 3.4.2 implies $z = 0$ a.e., hence $x = 0$ a.e. on E. □

The corresponding theorem for Y^*-valued X' uses Corollary A.2.4:

Theorem 3.4.11. *Let Y be a Banach space with the bidual property, and Y^* be its dual space. Let X be a pre-ideal space over some measure space S, and X' its Y^*-valued associate space. If $x : S \to Y$ is measurable with*

$$\int_S y(s)x(s)ds = 0 \qquad (y \in X',\ yx \geq 0),$$

then $x = 0$ a.e. on every subset of $\underline{\operatorname{supp}} X$ of σ-finite measure.

Proof. Let $E \subseteq \underline{\text{supp} X}$ have σ-finite measure. By Theorem 3.4.6 there is some $u \in X'$ with $\text{supp} u = E$. If we apply Corollary A.2.4 for $M = \{y \in X' : |y| = |u|\}$ we find that

$$\int_S |u(s)|\,|x(s)|\,ds = \sup_{y \in M} \int_S |y(s)|\,|x(s)|\,ds = \sup_{y \in M,\ yx \geq 0} \int_S y(s)x(s)ds.$$

The right hand side of the last equation vanishes by our assumption, hence so must do the left hand side, which implies that $x = 0$ a.e. on $\text{supp} u = E$. $\square$

3.5 Dual Spaces and Reflexivity

While in the previous section we were mainly interested in determining properties of integral functionals and characterizing functions that define such functionals, we will now examine, whether *any* functional on a pre-ideal space X is an integral functional. If this is the case, we write $X' = X^*$, where equality is meant in the sense of Definition 3.4.2. Recall (Theorem 3.4.4) that in this sense we always have $X' \subseteq X^*$ and $\|y\|_{X'} = \|y\|_{X^*}$.
Furthermore, to prove the equality $X' = X^*$ it suffices to show that for any $y^* \in X^*$ there exists some measurable y vanishing outside $\underline{\text{supp}X}$ and satisfying (3.16).

For Y-valued spaces X the problem is very hard. If Y is a Banach lattice, some useful sufficient conditions are given in [36]. But for general Y, there is no easy answer, not even for Lebesgue-Bochner spaces $X = L_p(S, Y)$ $(1 \le p < \infty)$: If S is an interval and Y has the so-called property (D) (which roughly means that all absolutely continuous functions $S \to Y$ satisfy FTC; see the remark in front of Lemma 4.4.1), it is shown in [7] that any bounded linear functional on X is of the form (3.16). A partial converse is proved, too (for reflexive X). For similar results, see [20]. A short survey can be found in [30].

For the rest of this section we will restrict ourselves to real-valued pre-ideal spaces X. We start with the classical result that $X' = X^*$ for regular X (over σ-finite measure spaces). But we will show even more that in Solovay's model this remains true without the regularity of X. In particular this means that $X' = X^*$ can not be disproved in Zermelo's set theory with the principle of dependent choices, but without the axiom of choice (provided Solovay's model exists).

However, without any assumptions on the underlying measure space, $X' = X^*$ is false, in general (this explains, why we will restrict ourselves to σ-finite measure spaces):

Example 3.5.1. Let all subsets of some set S be measurable and have infinite measure (except for the empty set). Then the associate space of $X = L_\infty(S, \mathbb{R})$ is trivial. But for any fixed $s \in S$ we may define a nontrivial bounded linear functional l on X by $l(x) = x(s)$.

The following fact is well-known:

Theorem 3.5.1. *Let X be a real-valued pre-ideal space over some σ-finite measure space. Then $l \in X^*$ is an integral functional if and only if for any sequence $x_n \in X$, $x_1 \ge x_2 \ge \ldots \ge 0$, $x_n \to 0$ a.e., we have $l(x_n) \to 0$.*

The proof may be found in [48, §69 Theorem 3]. It is based on the Radon-Nikodym theorem.

Since for regular X Theorem 3.3.5 implies $|l(x_n)| \leq \|l\| \, \|x_n\| \to 0$, we have:

Corollary 3.5.1. *$X' = X^*$ holds for any regular real-valued pre-ideal space X over some σ-finite measure space.*

Corollary 3.5.1 is the best possible result in this direction, if we assume the axiom of choice, as is shown in [48, §72 Theorem 5]:

Remark 3.5.1. If X is a real-valued pre-ideal space over some σ-finite measure space, and if Hahn-Banach's extension theorem for X holds true, then $X' = X^*$ implies that X is regular.

Corollaries 3.5.1 and 3.4.6 yield:

Corollary 3.5.2. *If X is a perfect real-valued ideal space over some σ-finite measure space, and if X and X' both are regular, then X is reflexive.*

Also Corollary 3.5.2 can not be sharpened, in general: In [48, §73 Theorem 2] it is shown (again using the axiom of choice) that any reflexive real-valued pre-ideal space X over some σ-finite measure space must necessarily be a regular perfect ideal space with regular X'.

However, we will prove now that both corollaries may be essentially sharpened in Solovay's model.

Definition 3.5.1. *Let X be a linear space of (classes of) measurable functions $S \to \mathbb{R}$. On the linear space of all linear functionals $l : X \to \mathbb{R}$ we define a partial ordering by $l \leq L$ if and only if*

$$l(x) \leq L(x) \qquad (0 \leq x \in X).$$

A linear functional $l : X \to \mathbb{R}$ is called positive functional *on X, if $l \geq 0$.*

In particular, for any positive linear functional l on Y, we have that $x \leq y$ implies $l(x) \leq l(y)$.
We will need the Jordan decomposition of linear functionals:

Theorem 3.5.2. *Let X be a real-valued pre-ideal space, and l a bounded linear functional on X. Then l is the difference of two positive bounded linear functionals. There even exists a decomposition $l = l_1 - l_2$, $0 \leq l_i \in X^*$ such that for any other decomposition $l = L_1 - L_2$, $0 \leq L_i \in X^*$, we have $L_1 - l_1 = L_2 - l_2 \geq 0$.*

The proof can be found in [48, §48 Theorem 1], although the theorem there is formulated only for perfect ideal spaces. You may also observe that X is a normed vector lattice, and thus X^* is a vector lattice, cf. proof of [41, II, Proposition 5.5] (or just use the statement there and apply [41, II, Proposition 1.4(3)]).

Definition 3.5.2. *A positive bounded linear functional L on a pre-ideal space X is called* singular functional, *if the inequality $0 \leq l \leq L$ for some integral functional $l \in X^*$ implies $l = 0$. An arbitrary $L \in X^*$ is called* singular functional, *if the positive functionals in its Jordan decomposition are singular.*

Theorem 3.5.3. *Let X be a real-valued pre-ideal space over some σ-finite measure space. Then the space X_i^* of all integral functionals of X and the space X_s^* of all singular functionals of X are closed linear subspaces of X^*, and X is the direct sum of X_i^* and X_s^*.*

The proof of the theorem may be found in [48, §70 Theorem 2 and subsequent remarks].

Theorem 3.5.4. *Let in Theorem 3.5.3 the positive linear functional $l \in X^*$ have the (unique) representation $l = l_i + l_s$ with $l_i \in X_i^*$, $l_s \in X_s^*$, then $l_i, l_s \geq 0$, and for $x \geq 0$ we have that $l_i(x)$ is the infimum of all $\lim l(x_n)$, where $x_n \geq 0$ converges a.e. monotonically increasing to x.*

For the proof see [48, §70 Theorem 2 and Exercise 70.1].

Recall the known 'constructions' of singular functionals l on $X = L_\infty(S)$, when S is a Borel space (for the notation see Definition 4.1.4): One typically fixes some essential accumulation point s_0 of S and puts $l(x) = \operatorname{ess\,lim}_{s \to s_0} x(s)$ on the linear subspace of all x, where this limit exists. Then Hahn-Banach's theorem is applied. The next theorem shows that this 'construction' is the only one in some sense:

Theorem 3.5.5. *Let X be a real-valued pre-ideal space over some σ-finite measure space. Let $l \in X^*$ be a positive singular functional. Then to any $0 \leq x \in X$ there exists a sequence $x_n \leq x$ converging a.e. monotonically decreasing to 0, such that $l(x_n) = l(x)$.*

Proof. We equip X with the different seminorm $\|x\| = l(|x|)$ and recall Remark 3.2.3: Proposition 3.2.1 shows us that the Lorentz seminorm $\|\cdot\|_L$ satisfies $\|x\|_L = l_i(x)$ for any $0 \leq x \in X$, where $l = l_i + l_s$ as in Theorem 3.5.4. Now $l_i = 0$ implies $\|x\|_L = 0$. Corollary 3.2.6 thus yields that there exists a sequence $y_n \geq 0$ converging a.e. monotonically increasing to x with $\|y_n\| \to \|x\|_L$, i.e. $l(y_n) \to 0$. Hence $l(y_n) = 0$, since l is positive. Now just put $x_n = x - y_n$. □

By Hahn-Banach's theorem there do exist singular functionals. Since we want to prove that in some model of set theory (in which the principle of dependent choices is true) there do not exist such functionals, Hahn-Banach's theorem (and thus also the axiom of choice) must fail in this model. We recall that Hahn-Banach's theorem is equivalent to the fact that to any Boolean algebra

B there exists a nonnegative function μ on B such that $\mu(0) = 0$, $\mu(1) = 1$ and $\mu(a+b) = \mu(a) + \mu(b)$ whenever $a \cdot b = 0$, see [34] (cf. also [24, Problem 2.6.19]). In particular, Hahn-Banach fails if the following proposition is true:

(PM$_\omega$) There is no finitely additive probability measure defined on the power set of natural numbers which vanishes on singletons.

This proposition is satisfied e.g., when it is interpreted in Solovay's model of set theory, as was already remarked in [44, p. 3]; for a proof see [37]. Thus (provided Solovay's model exists), it will not lead to a contradiction to assume that (PM$_\omega$) is true. For a detailed discussion of the consistency of (PM$_\omega$) we refer the reader to [37]. The following theorem shows that (PM$_\omega$) already implies that $X' = X^*$:

Theorem 3.5.6. *Let X be a real-valued pre-ideal space over some σ-finite measure space with $X' \neq X^*$. Then (PM$_\omega$) is false, i.e. there exists a function $\mu : \mathcal{P}(\mathbb{N}) \to [0,1]$ with $\mu(\mathbb{N}) = 1$, $\mu(M \cup N) = \mu(M) + \mu(N)$ for disjoint $M, N \subseteq \mathbb{N}$, and $\mu(\{n\}) = 0$ for all $n \in \mathbb{N}$.*

Proof. Since the space of integral functionals on X is a linear subspace by Theorem 3.5.3, Theorem 3.5.2 implies that there exists a positive bounded linear functional on X, which is not an integral functional. Thus, by Theorem 3.5.4, there exists a nontrivial positive singular functional $l \in X^*$.
Since l is nontrivial there exists some $x \in X$ with $l(x) = 1$, and by Theorem 3.5.5 even a sequence $x_n \geq 0$ converging a.e. monotonically decreasing to 0 with $l(x_n) = 1$. Now put $y_n = x_n - x_{n+1}$. Then $y_n \geq 0$, $\sum y_n = x_1$, whence for any sequence α_n taking only the values 0 and 1, we have that $y = \sum \alpha_n y_n$ satisfies $0 \leq y \leq x$, and thus $y \in X$, $0 \leq l(y) \leq l(x_1) = 1$. For $M \subseteq \mathbb{N}$ we put

$$\alpha_n(M) = \begin{cases} 1 & n \in M, \\ 0 & n \notin M \end{cases}$$

and define

$$\mu(M) := l(\sum \alpha_n(M) y_n).$$

Then $\mu(\mathbb{N}) = l(x_1) = 1$, $\mu(\{n\}) = l(y_n) = l(x_n) - l(x_{n+1}) = 0$, and for disjoint $M, N \subseteq \mathbb{N}$, we have $\alpha_n(M \cup N) = \alpha_n(M) + \alpha_n(N)$, whence

$$\mu(M \cup N) = l(\sum (\alpha_n(M) + \alpha_n(N)) y_n) = \mu(M) + \mu(N).$$

Thus μ has the desired properties. □

The function μ constructed in the proof of Theorem 3.5.6 may take other values than 0 and 1, if we allow the application of Hahn-Banach:

Example 3.5.2. Let $X = l_\infty$, and define $l(x) = \lim x_n$ on the subspace U of all $x = (x_n)_n \in X$ for which the limit exists. Choose $y = (y_n)_n \in X$ with $y_n \in \{0, 1\}$ such that $\lim y_n$ does not exist, $0 < \alpha < 1$, and put $l(y) = \alpha$. By the formula $l(x + \lambda y) = l(x) + \lambda l(y)$, l extends to a linear functional on the linear span V of $U \cup \{y\}$.
l is bounded on V by 1, since for any $x = (x_n)_n \in U$ and any λ it is easily checked that $\|x + \lambda y\| \geq |\lim x_n|$ and $\|x + \lambda y\| \geq |\lim x_n + \lambda|$, and thus $|l(x + \lambda y)| = |\lim x_n + \alpha\lambda| \leq \|x + \lambda y\|$. Similarly, l is positve: Indeed, $x + \lambda y \geq 0$ for $x = (x_n)_n \in U$ implies $\lim x_n \geq 0$ and $\lim x_n + \lambda \geq 0$, whence also $l(x + \lambda y) = \lim x_n + \alpha\lambda \geq 0$.
V is a vector sublattice of X (for the notation see [41]): We have to show that for any $x = (x_n)_n, z = (z_n)_n \in U$ and any $\lambda, \mu \in \mathbb{R}$ we have that $s = \sup(x + \lambda y, z + \mu y) \in V$. Obviously, $s = (s_n)_n$ with $s_n = a_n + (b_n - a_n)y_n$, where we have put $a_n = \max\{x_n, y_n\}$, $b_n = \max\{x_n + \lambda, z_n + \nu\}$. Since $a_n \to a$ and $b_n \to b$ converge, we may conclude that $s_n - (b - a)y_n \to a$ converges, i.e. $s \in U$.
Thus, Hahn-Banach's theorem for normed vector lattices [41, II, Proposition 5.6] implies that there exists a positive linear extension of l to an element of X^* with norm 1. This extension (again denoted by l) is a singular functional: Given any $0 \leq x = (x_k)_k \in X$ consider the sequence $y_n = (y_k^n)_k \in X$, defined by

$$y_k^n = \begin{cases} x_k & k \leq n, \\ 0 & k > n. \end{cases}$$

Then y_n converges monotonically increasing to x, but $l(y_n) = 0$. This means that the integral part l_i of l vanishes by Theorem 3.5.4, i.e. l is singular.
Now we consider the proof of Theorem 3.5.6 for this l and for $x_n = (x_k^n)_k$, where

$$x_k^n = \begin{cases} 1 & k \geq n, \\ 0 & k < n \end{cases}$$

(you may choose $x = x_1$). Then $\mu(N) = l(\chi_N)$, and thus μ takes also the value $l(y) = \alpha$.

Observing that $l_\infty^* = l_1$ contradicts Hahn-Banach and recalling the remark at the introduction of (PM$_\omega$), it does not seem very likely that already a weaker assumption than (PM$_\omega$) implies $X^* = X'$. Example 3.5.2 shows that – at least with our method of proof – it is not possible to weaken (PM$_\omega$) to:

(UF$_\omega$) There is no finitely additive probability measure μ defined on the power set of natural numbers which vanishes on singletons and takes only the values 0 and 1.

Observe that (UF$_\omega$) is equivalent to the fact that there exists no ultrafilter $\mathcal{F}$ on $\mathbb{N}$ containing the co-finite sets: Indeed, given some such filter $\mathcal{F}$, it is easily checked that $\mu(M) = 1$ for $M \in \mathcal{F}$ and $\mu(M) = 0$ for $M \notin \mathcal{F}$ gives a measure contradicting (UF$_\omega$); conversely, given such a measure μ,

and defining $\mathcal{F}$ by the above relations, a straightforward calculation shows that $\mathcal{F}$ is an ultrafilter on $\mathbb{N}$ containing the co-finite sets.
The assumption (UF_ω) is very weak, since its failing already implies the existence of subsets of the real line, which are not Lebesgue measurable [43] (see also [24, Problem 1.4.10]). You may also observe that, if there exists an ultrafilter $\mathcal{F}$ on $\mathbb{N}$ containing the co-finite sets, it is δ-incomplete. Thus there exists a higher order nonstandard model of $\hat{\mathbb{R}}$, whence also a nonmeasurable function on $[0,1]$, see [35] (cf. also [32] and [45]). In contrast to this, according to [37], it is still an open problem, whether also the failing of (PM_ω) implies the existence of a nonmeasurable subset of the real line.

We now discuss some consequences of Theorem 3.5.6. Observing that $X^* = X'$ and $(X')^* = (X')'$ implies $X^{**} = X''$, Corollary 3.4.6 yields:

Corollary 3.5.3. *Let (PM_ω) hold. Then a real-valued pre-ideal space over some σ-finite measure space is reflexive if and only if it is perfect.*

In particular, under the assumption (PM_ω), the spaces L_1, L_∞, l_1, l_∞ all are reflexive!

Theorem 3.4.8 implies by the same argument:

Corollary 3.5.4. *Let (PM_ω) hold. Then a real-valued pre-ideal space over some σ-finite measure space has the bidual property if and only if it is almost perfect.*

Recalling that Hahn-Banach holds in separable spaces we also have in view of Remark 3.5.1:

Corollary 3.5.5. *Let (PM_ω) hold. If a real-valued pre-ideal space over some σ-finite measure space is not regular, then it is not separable.*

In particular, assuming (PM_ω) it may happen that a Banach space X is not separable although X^* is! (Choose e.g. $X = l_\infty$).

We remark that Corollary 3.5.5 also holds without the assumption (PM_ω). This is proved in [50].

4. Ideal Spaces on Product Measures and Calculus

4.1 Spaces with Mixed Norm

In this section, T and S will always denote σ-finite measure spaces.

Definition 4.1.1. *A normed linear space U of (classes of) measurable functions over S is called* T-measurable, *if for each product-measurable x on $T \times S$, satisfying $x(t, \cdot) \in U$ for almost all $t \in T$, we have that $t \mapsto \|x(t, \cdot)\|_U$ is measurable.* S-measurability *is defined analogously.*

Using the axiom of choice, it is shown in [32] that there *do* exist natural measure spaces T and S and an (even semi-perfect) ideal space U over S, such that U is not T-measurable.

However, natural ideal spaces usually *are* T-measurable, although this sometimes might be very hard to prove. The most important fact for applications in this connection is the Luxemburg-Gribanov Theorem A.3.1:

Theorem 4.1.1. *Let U be a pre-ideal space over S. Then for any measurable function x on $T \times S$ the function $t \mapsto \|x(t, \cdot)\|_{U'}$ is measurable.*

Proof. Put $B = \{|u| : u \in U,\ \|u\|_U \leq 1\}$ in Theorem A.3.1. $\square$

Corollary 4.1.1. *Each perfect ideal space U over S is T-measurable.*

Proof. $U = U''$ with the same norm by Corollary 3.4.6. Apply Theorem 4.1.1 for $\tilde{U} = U'$. $\square$

Corollary 4.1.1 was first proved in [31]. In [32] it was suspected that perfectness might even be necessary for U to be T-measurable. We will see soon that this is by far not the case: There do exist many 'natural' counterexamples. Thus it is worth studying different approaches to T-measurability. We start with a simple lemma of measure theory (which holds on arbitrary measure spaces):

Lemma 4.1.1. *Let x be a measurable function with σ-finite support. Then there exists a sequence of simple functions x_n, which converges to x a.e., and additionally satisfies $|x_1| \leq |x_2| \leq \ldots \leq |x|$ a.e.*

Proof. Since supp$x$ is σ-finite, there exists a sequence of simple functions y_n, which converges to x a.e. Without loss of generality we may assume that supp$x$ = suppy_n. For the same reason there exists a sequence of nonnegative simple functions α_n, which converges to $|x|$, and additionally satisfies $\alpha_n \leq |x|$. Now define $\beta_n = \max_{k \leq n} \alpha_k$. Then β_n converges monotonically increasing to $|x|$. Now just define $x_n(s) = \beta_n(s) \, |y_n(s)|^{-1} \, y_n(s)$ (for $y_n(s) = 0$ let $x_n(s) = 0$). □

This has the pleasant consequence that in the most important cases T-measurability is simpler to check.

Definition 4.1.2. *A normed linear space U of measurable functions over S is called* T-simple measurable, *if for each simple function x on $T \times S$, satisfying $x(t, \cdot) \in U$ for almost all $t \in T$, we have that $t \mapsto \|x(t, \cdot)\|_U$ is measurable.*

Lemma 4.1.1 almost immediately implies the

Lemma 4.1.2. *Let U be an almost perfect pre-ideal space over S. Then U is T-measurable, if and only if U is T-simple measurable.*

Proof. Let x be measurable with $x(t, \cdot) \in U$ for almost all t. Let x_n be the sequence of Lemma 4.1.1. By Corollary 3.2.5 we have $\|x_n(t, \cdot)\|_U \to \|x(t, \cdot)\|_U$ for almost all t. □

We will now formulate another sufficient condition for T-measurability:

Definition 4.1.3. *A pre-ideal space U has* simple-regular support, *if it has the following property:*
Whenever a set E has finite measure, such that χ_E belongs to the real form of U, then χ_E has absolute continuous norm.

Theorem 4.1.2. *Let U be an almost perfect pre-ideal space over S with simple-regular support. Then U is T-measurable.*

The theorem will be a consequence of Theorem 4.3.2. Together with Corollary 3.3.4 we have the

Corollary 4.1.2. *Let U be a regular pre-ideal space over S. Then U is T-measurable.*

In the example of Orlicz spaces generated by finite Young functions (which are not regular, in general), we will show now that Theorem 4.1.2 is applicable in more cases than Corollary 4.1.2. Of course, the conclusion that such Orlicz spaces are T-measurable, is just of little interest, since this result already is a special case of Corollary 4.1.1. But we will see in the next section that the following reasoning may also be used together with a generalization of

Theorem 4.1.2 to gain a result on T-measurability of some classes of families of such Orlicz spaces (Definition 4.3.1), for which a generalization of Corollary 4.1.1 is not known.

Lemma 4.1.3. *Let $(X, \|\cdot\|)$ be a pre-ideal space with simple-regular support. Let $\|\cdot\|^*$ be another norm on X, which turns X into a pre-ideal space and satisfies*

$$\|x\|^* \leq c\,\|x\| \qquad (x \in X)$$

for some $c > 0$. Then also $(X, \|\cdot\|^*)$ has simple-regular support.

Proof. Let $x = \chi_E \in X$, where E has finite measure. Then $\|P_D x\|^* \leq c\,\|P_D x\| \to 0$ for $\text{mes}D \to 0$. □

Theorem 4.1.3. *If $U = L_\Phi(S)$ is an (ideal) Orlicz space generated by a finite Young function Φ (with the Luxemburg norm or with the Orlicz norm), then U has simple-regular support.*

Proof. First consider the Luxemburg norm on X, i.e.

$$\|x\| = \inf\{\alpha > 0 : \int_S \Phi\left(\frac{|x(s)|}{\alpha}\right) ds \leq 1\}.$$

Let $x = \chi_E \in U$ with $\text{mes}E < \infty$. We have to prove that $\|P_D x\| \to 0$ for $\text{mes}D \to 0$. But for any $\varepsilon > 0$ we have

$$\int_S \Phi\left(\frac{|P_D x(s)|}{\varepsilon}\right) ds = \Phi(\varepsilon^{-1})\text{mes}(D \cap E) \leq 1$$

for $\text{mes}D$ small enough, whence $\|P_D x\| \leq \varepsilon$. The corresponding statement for the Orlicz norm follows by Lemma 4.1.3, since that norm is equivalent to the Luxemburg norm (see e.g. [39]). □

The conditions of Corollary 4.1.1 and Theorem 4.1.2 are by no means necessary for T-measurability. In fact, we will show now that the most typical examples of not almost perfect pre-ideal spaces are all T-measurable.

We prepare the result by a lemma, which for nonnegative x is a special case of the fact that $L_\infty(S)$ is T-measurable (Corollary 4.1.1). However, we give an elementary proof:

Lemma 4.1.4. *If $x : T \times S \to [-\infty, \infty]$ is measurable, then*

$$y(t) = \operatorname*{ess\,sup}_{s \in S} x(t, s)$$

is measurable.

Proof. We have to prove that $M_c = \{t : \operatorname{ess\,sup}_s x(t,s) > c\}$ belongs to the Lebesgue-extension of the measure for any c. Let

$$x_c(t,s) = \begin{cases} 1 & \text{if } x(t,s) > c, \\ 0 & \text{if } x(t,s) \leq c. \end{cases}$$

Then $t \in M_c$, if and only if $x_c(t,\cdot)$ is no null function. Since x_c is measurable and nonnegative, the theorem of Fubini-Tonelli implies that $y_c(t) = \int_S x_c(t,s)ds$ is defined a.e. and measurable. Thus, since M_c and $\{t : y_c(t) > 0\}$ differ at most by null sets, we are done. □

To cover a quite general case with our theorem, we recall that a Borel space is a topological space S with measurable sets generated by the open sets. We call a point s_0 with a countable base of neighborhoods an *essential accumulation point* of S, if for any null set $N \subseteq S$ there exists at least one sequence $s_n \in S \setminus N$, $s_n \neq s_0$, which converges to s_0.

Definition 4.1.4. *Let S be a Borel space, s_0 be an essential accumulation point of S, and $x : S \to [-\infty, \infty]$. Then we define*

$$\begin{aligned} \operatorname*{ess\,lim\,sup}_{s \to s_0} x(s) &:= \inf_{\operatorname{mes} N = 0} \sup_{\substack{s_n \to s_0 \\ s_0 \neq s_n \notin N}} \limsup_{n \to \infty} x(s_n) \\ \operatorname*{ess\,lim\,inf}_{s \to s_0} x(s) &:= \sup_{\operatorname{mes} N = 0} \inf_{\substack{s_n \to s_0 \\ s_0 \neq s_n \notin N}} \liminf_{n \to \infty} x(s_n). \end{aligned}$$

If ess lim sup *and* ess lim inf *coincide, we write* ess lim *instead.*

We emphasize that in cases like $S = \mathbb{C}$, $S = \mathbb{R}$ or $S = \mathbb{N}$ the point $s_0 = \infty$ (resp. $s_0 = -\infty$) is easily included with its usual system of neighborhoods, such that the following Theorem 4.1.4 is also applicable in that cases.
In case $S = \mathbb{N}$, $s_0 = \infty$ we of course have that $\operatorname{ess\,lim\,sup} x(s) = \limsup x(n)$, such that the ideal space of sequences $x = (x_n)_n$, generated by the norm

$$\|x\| = \sup |x_n| + \limsup_{n \to \infty} |x_n|,$$

is a special case of the following class of pre-ideal spaces:
Let X be the space generated by the norm

$$\|x\| = \|x\|_{L_p(S)} + \sup_{s_0 \in H} a_{s_0} \operatorname*{ess\,lim\,sup}_{s \to s_0} |x(s)|,$$

where $1 \leq p \leq \infty$, H is a subset of essential accumulation points of the underlying Borel space S, and $a_{s_0} > 0$. Then usually X is neither almost perfect nor regular, but T-measurable, if H is at most countable by the following Theorem 4.1.4. We remark that the previous example by far does not cover all examples of not almost perfect pre-ideal spaces: For example, instead of the weighted supremum one could choose some other norm of a pre-ideal space over H (if H is measurable).

Theorem 4.1.4. *Let S be a σ-finite Borel space, and s_0 be an essential accumulation point of S, such that $\{s_0\}$ is measurable. Let T be some σ-finite measure space. Then for any measurable $x : T \times S \to [-\infty, \infty]$ the functions*

$$y(t) = \operatorname{ess}\limsup_{s \to s_0} x(t,s) \quad \text{and} \quad z(t) = \operatorname{ess}\liminf_{s \to s_0} x(t,s)$$

are measurable.

Proof. Since $z(t) = -\operatorname{ess}\limsup_{s \to s_0} -x(t,s)$, it suffices to prove that y is measurable. Choose a countable base U_n of open neighborhoods of s_0 with $U_1 \supseteq U_2 \supseteq \dots$. At first we state that for any null set N and any measurable $f : S \to [-\infty, \infty]$ we have

$$L(N) := \inf_n \sup_{s_0 \neq s \in U_n \setminus N} f(s) = \sup_{\substack{s_n \to s_0 \\ s_0 \neq s_n \in U_n \setminus N}} \limsup_{n \to \infty} f(s_n) =: R(N). \tag{4.1}$$

In fact, if for any n we choose some $s_n \in U_n \setminus N$, $s_n \neq s_0$ with $f(s_n) \geq \sup_{s_0 \neq s \in U_n \setminus N} f(s) - n^{-1}$ we have $s_n \to s_0$ by $U_1 \supseteq U_2 \supseteq \dots$, and thus $L(N) \leq \limsup f(s_n) \leq R(N)$. On the other hand, let $s_n \to s_0$ with $s_0 \neq s_n \notin N$ be arbitrary, and $G = \limsup f(s_n)$. Then for each fixed n almost all s_k belong to U_n, i.e. $G \leq \sup_{s_0 \neq s \in U_n \setminus N} f(s)$. Consequently, $G \leq L(N)$. Since s_n was arbitrary this implies $R(N) \leq L(N)$, and (4.1) is proved. Now we have

$$\begin{aligned} \operatorname{ess}\limsup_{s \to s_0} f(s) &= \inf_{\operatorname{mes} N = 0} R(N) = \inf_{\operatorname{mes} N = 0} L(N) \\ &= \inf_n \inf_{\operatorname{mes} N = 0} \sup_{s_0 \neq s \in U_n \setminus N} f(s) = \inf_n \operatorname*{ess\,sup}_{s \in U_n \setminus \{s_0\}} f(s). \end{aligned}$$

This means

$$y(t) = \inf_n \operatorname*{ess\,sup}_{s \in U_n \setminus \{s_0\}} x(t,s),$$

and thus y is measurable by Lemma 4.1.4. $\square$

For subsets S of the (extended) real line the previous lemma implies also that functions like

$$y(t) = \operatorname{ess}\limsup_{s \to s_0^-,\, s \in S} x(t,s) \quad \text{or} \quad y(t) = \operatorname{ess}\limsup_{s \to s_0^+,\, s \in S} x(t,s)$$

(which are defined in the obvious way) are measurable: To see this, just apply the lemma for $S \cap [-\infty, s_0]$ resp. $S \cap [s_0, \infty]$ instead of S. Thus, we have even more examples of T-measurable but not almost perfect pre-ideal spaces.

The T-measurability ensures that the spaces with mixed norms are in fact normed linear spaces:

Definition 4.1.5. *If U is T-measurable and V is a pre-ideal space over T, then the space $[U \to V]$ with* mixed norm *consists of all product-measurable functions x on $T \times S$, for which the function $t \mapsto \|x(t,\cdot)\|_U$ is defined a.e. and belongs to V. The norm is defined by*

$$\|x\|_{[U \to V]} = \|t \mapsto \|x(t,\cdot)\|_U\|_V .$$

Similarly, if V is S-measurable, and U is a pre-ideal space over S, the space $[U \leftarrow V]$ is defined analogously by the norm

$$\|x\|_{[U \leftarrow V]} = \|s \mapsto \|x(\cdot,s)\|_V\|_U .$$

In the following, U and V will always denote pre-ideal spaces over the measure-spaces S and T resp.

Then $[U \to V]$ and $[U \leftarrow V]$ are pre-ideal (if defined).

The following theorem is not as trivial as you might suspect, since a set $E \subseteq T \times S$ of positive measure does not necessarily contain a set of the form $A \times B$ of positive measure, even if you are allowed to change E on null sets (see Example 4.3.1).

Theorem 4.1.5. *Let U and V be pre-ideal spaces. If $[U \to V]$ or $[U \leftarrow V]$ are defined, their support is* $\operatorname{supp} V \times \operatorname{supp} U$.

Proof. We prove that $X = [U \to V]$ has support $Q = \operatorname{supp} V \times \operatorname{supp} U$. Let $x \in X$. Since $x(t,\cdot) \in U$, $x(t,s)$ vanishes for almost all (t,s) outside $T \times \operatorname{supp} U$. Similarly, we have that $\|x(t,\cdot)\|$ vanishes for almost all t outside $\operatorname{supp} V$, hence $x(t,s) = 0$ for almost all (t,s) outside $\operatorname{supp} V \times S$. Together we have $x(t,s) = 0$ for almost all (t,s) outside Q. For the converse, Corollary 2.2.7 yields that there are functions $u_n \in U$, $v_n \in V$ with $\bigcup \operatorname{supp} u_n = \operatorname{supp} U$, $\bigcup \operatorname{supp} v_n = \operatorname{supp} V$, $\operatorname{supp} u_1 \subseteq \operatorname{supp} u_2 \subseteq \ldots$, and $\operatorname{supp} v_1 \subseteq \operatorname{supp} v_2 \subseteq \ldots$. Then the sequence $x_n(t,s) = |v_n(t)|\, u_n(s)$ lies in X, and by $\operatorname{supp} x_n = \operatorname{supp} u_n \times \operatorname{supp} v_n$ we have $\bigcup \operatorname{supp} x_n = Q$. Thus, if $E \subseteq Q$ has positive measure, we have by $E = \bigcup (E \cap \operatorname{supp} x_n)$ that at least one of the sets $E \cap \operatorname{supp} x_n$ must have positive measure. □

The following theorems will be proved later in a more general setting.

Theorem 4.1.6. *If U and V are semi-perfect pre-ideal spaces, then $[U \to V]$ and $[U \leftarrow V]$ are semi-perfect ideal spaces (if defined).*

Theorem 4.1.7. *Let U be an almost α-perfect pre-ideal space, and V be an almost β-perfect pre-ideal space. Then $[U \to V]$ and $[U \leftarrow V]$ are almost $\alpha\beta$-perfect pre-ideal spaces (if defined).*

In view of Corollary 4.1.1 we have:

Corollary 4.1.3. *If U and V are perfect ideal spaces, then $[U \to V]$ and $[U \leftarrow V]$ are defined and perfect ideal spaces.*

Theorem 4.1.8. *If U and V are regular pre-ideal spaces, then $[U \to V]$ and $[U \leftarrow V]$ are defined and regular pre-ideal spaces.*

Theorem 4.1.9. *If $[U \to V]$ or $[U \leftarrow V]$ is a nontrivial pre-ideal space, then so is U and V. If additionally $[U \to V]$ or $[U \leftarrow V]$ is complete, almost α-perfect, semi-perfect, or regular, then U and V have this property, too.*

Theorem 4.1.10. $[U \to V]' = [U' \to V']$ *and* $[U \leftarrow V]' = [U' \leftarrow V']$ *with the same norm (provided that $[U \to V]$ resp. $[U \leftarrow V]$ are defined).*

Observe that by Theorem 4.1.1 the spaces $[U' \to V']$ and $[U' \leftarrow V']$ are always defined.

We remark that for perfect ideal spaces U and V, since $[U \to V]$ is perfect (Corollary 4.1.3), Corollary 3.4.6 implies that Theorem 4.1.10 is equivalent to $[U \to V] = [U' \to V']'$ (with the same norms). For $U = L_p(S)$, $V = L_q(T)$ it is proved in [6] that

$$\|x\|_{[U \to V]} = \|x\|_{[U' \to V']'} \qquad (x \in [U \to V]).$$

But, as was emphasized by Chur-jen Chen 1996, you can not conclude from this proof that $[U \to V] = [U' \to V']'$ (i.e. that Theorem 4.1.10 holds for this case), since the assumption $x \in [U \to V]$ is essential there. All you can conclude is that $Y = [U' \to V']$ $(= [U' \to V']'')$ is continuously embedded in $X' = [U \to V]'$ with embedding constant 1, and that *if* Y and X' coincide as sets, then they have the same norm (because then the sets $[U' \to V']'$ and $[U \to V]'' = [U \to V]$ coincide in view of Corollary 3.4.1).
That the first fact is true in much more general cases will be proved in Lemma 4.3.6. But the inclusion $X' \subseteq Y$ is much deeper, and we will use the Luxemburg-Gribanov Theorem A.3.2 for its proof in Theorem 4.3.8.

4.2 Weighted Spaces and Projections of Spaces

Definition 4.2.1. *If X is a normed linear space of (classes of) measurable functions, and w is some measurable scalar weighting function, then the* weighted function space *$X(w)$ consists of all measurable functions x with* $\operatorname{supp} x \subseteq \operatorname{supp} w$, *such that $wx \in X$. The norm is defined by $\|x\|_{X(w)} = \|wx\|_X$.*

Theorem 4.2.1. *If X is a pre-ideal space, then $X(w)$ is a pre-ideal space. If additionally X is almost α-perfect, semi-perfect, inner-regular, outer-regular, or complete, then $X(w)$ has the same property.*

Proof. We prove completeness. Let x_n be Cauchy in $X(w)$. Then wx_n is Cauchy in X, i.e. $wx_n \to y$ in X. Lemma 2.2.5 yields $\operatorname{supp} y \subseteq \operatorname{supp} w$. Thus there exists a function x with $\operatorname{supp} x \subseteq \operatorname{supp} w$ and $y = wx$ a.e. Then $x \in X(w)$, $\|x_n - x\|_{X(w)} = \|wx_n - y\|_X \to 0$. □

We now define the converse of the construction of spaces with mixed norms. For the rest of this section let T and S be σ-finite measure spaces. A function space X over $T \times S$ may be used to define a function space over S:

Definition 4.2.2. *Let X be a normed linear space of (classes of) $T \times S$-measurable functions, and w be a product-measurable scalar weighting function. Then the* weighted projection *$X|_S^w$ consists of all measurable functions x on S with the following two properties:*

(a) The function $y(t,s) = w(t,s)x(s)$ belongs to X, and
(b) for almost all s, for which $w(\cdot, s)$ vanishes a.e., we have $x(s) = 0$.

The norm is defined by $\|x\|_{X|_S^w} = \|y\|_X$.
The weighted projection *$X|_T^w$ is defined analogously.*

The projection may in fact be used to regain U and V from $[U \to V]$ or from $[U \leftarrow V]$:

Remark 4.2.1. If $u \in U$ and $v \in V$ are not identically vanishing, then

$$\begin{aligned} [U \to V]|_S^{(t,s)\mapsto v(t)} = [U \leftarrow V]|_S^{(t,s)\mapsto v(t)} &= U, \text{ and} \\ [U \to V]|_T^{(t,s)\mapsto u(s)} = [U \leftarrow V]|_T^{(t,s)\mapsto u(s)} &= V, \end{aligned}$$

where also the norms are equal up to the constant factor $\|v\|_V$ resp. $\|u\|_U$.

Since one goal of the following theory will be the application to integral operators (Lemma 5.3.1 and Theorem 5.3.1), where w is the kernel function of a linear integral operator, we do not want to exclude the case that w has zeros. This is the reason, why we use weighted projections instead of unweighted projections of weighted function spaces. If w has zeros the latter is more restrictive. We may see this for the kernel of a Volterra operator:

Example 4.2.1. Let $T = S = [0,1]$, $w(t,s) = 0$ for $t > s$, non-vanishing otherwise. Then $L_1(w)$ has no nontrivial unweighted projection, while $L_1|_S^w$ and $L_1|_T^w$ and even $L_1(w)|_S^w$ and $L_1(w)|_T^w$ are all nontrivial, if w is bounded.

Lemma 4.2.1. *Let x_n be measurable on S, and w be measurable on $T \times S$, such that $y_n(t,s) = w(t,s)x_n(s)$ converges to some y a.e. on the product space. Then $y(t,s) = w(t,s)x(s)$ for some measurable x. We may choose x in such a way that $x(s) = 0$, if $w(\cdot,s) = 0$ a.e.*

Proof. For almost all (t,s) with the additional assumption $w(t,s) \neq 0$ we have $x_n(s) \to w(t,s)^{-1}y(t,s)$. This shows that the right hand side is independent of t for almost all (t,s) for which $w(t,s) \neq 0$. Thus for almost all s with $w(\cdot,s) \neq 0$ we have $w(t,s)^{-1}y(t,s) = x(s)$ for some measurable x. Define $x(s) = 0$ for the remaining s. □

Lemma 4.2.2. *Let X be a pre-ideal* space over $T \times S$, and x lie in the regular part of X. Then for any $\varepsilon > 0$ there exist sets $E_T \subseteq T$, $E_S \subseteq S$ of finite measure, such that*

$$\|P_{(T\times S)\setminus(E_T\times E_S)}x\| < \varepsilon.$$

Proof. Since T, S are σ-finite, there exist increasing sequences $T_1 \subseteq T_2 \subseteq \ldots$, $S_1 \subseteq S_2 \subseteq \ldots$ of sets of finite measure with $T = \bigcup T_n$, $S = \bigcup S_n$. Then $Q_n = T_n \times S_n$ satisfies $Q_1 \subseteq Q_2 \subseteq \ldots$ and $T \times S = \bigcup Q_n$. Use Lemma 3.3.1. □

Lemma 4.2.3. *Let X be a pre-ideal* space over $T \times S$, and x lie in the regular part of X. Then*

$$\lim_{\delta\to 0} \sup_{\mathrm{mes} D \leq \delta} \|P_{T\times D}x\| = 0, \quad \text{and} \quad \inf_{\mathrm{mes} E < \infty} \|P_{T\times(S\setminus E)}x\| = 0.$$

Proof. Given $\varepsilon > 0$, choose E_T, E_S as in Lemma 4.2.2. Since

$$|\chi_{T\times D}| \leq |\chi_{(T\times S)\setminus(E_T\times E_S)} + \chi_{E_T\times D}|,$$

we have

$$\|P_{T\times D}x\| \leq \|P_{(T\times S)\setminus(E_T\times E_S)}x + P_{E_T\times D}x\| \leq 2\varepsilon,$$

if $\mathrm{mes} D$ is small enough, because $\mathrm{mes}(E_T \times D) \to 0$ for $\mathrm{mes} D \to 0$. The second statement follows by

$$|\chi_{T\times(S\setminus E_S)}| \leq |\chi_{(T\times S)\setminus(E_T\times E_S)}|.$$

□

In the proof of the following theorem we only show the nontrivial parts:

Theorem 4.2.2. *Let X be a pre-ideal* space over $T \times S$, w measurable on $T \times S$. Then $X|_S^w$ and $X|_T^w$ are pre-ideal* spaces. If X additionally is almost α-perfect, semi-perfect, regular, or if X is even a pre-ideal or an ideal space, then $X|_S^w$ and $X|_T^w$ have this property, too.*

Proof. To show that $X|_S^w$ is regular for regular X, use Lemma 4.2.3.
Let X be an ideal space. We show completeness of $X|_S^w$. Let x_n be Cauchy in $X|_S^w$. Then $y_n(t,s) = w(t,s)x_n(s)$ is Cauchy in X, hence it converges to some y in X. By Corollary 3.2.1 there exists a subsequence with $y_{n_k} \to y$ a.e. in the product-measure. Choose x as in Lemma 4.2.1. Then $x \in X$, $\|x_n - x\|_{X|_S^w} = \|y_n - y\|_X \to 0$. □

Since by Remark 4.2.1, U and V are weighted projections of $[U \to V]$ and $[U \leftarrow V]$ (if the spaces are nontrivial), Theorem 4.1.9 is just a special case of Theorem 4.2.2.

4.3 Spaces with Mixed Family Norm

In this section, T and S will denote σ-finite measure spaces.
Usually, $[U \to V](w)$ are not spaces with mixed norm over $T \times S$. However, they may be regarded as spaces with mixed family-norm:

In the next definition we consider instead of one fixed space U over S a family $U(\cdot)$ of such spaces:

Definition 4.3.1. *Let T, S be σ-finite measure spaces, and $U(t)$ for almost all $t \in T$ be a normed linear space of (classes of) measurable functions over S. We will call the family $U(\cdot)$* T-measurable, *if for any $T \times S$-measurable function x, satisfying $x(t, \cdot) \in U(t)$ for almost all t, we have that $t \mapsto \|x(t, \cdot)\|_{U(t)}$ is measurable.*

Corollary 4.1.1 and Corollary 4.1.2 ensure, for example, that any constant family of perfect or regular pre-ideal spaces is T-measurable.

The T-measurability ensures that the spaces described in the following definition are in fact normed linear spaces:

Definition 4.3.2. *Let the family $U(\cdot)$ be T-measurable, and V be a pre-ideal space over T. Then the* space with mixed family-norm $[U(\cdot) \to V]$ *consists of all $T \times S$-measurable functions x, for which the function $t \mapsto \|x(t, \cdot)\|_{U(t)}$ is defined a.e. and belongs to V. The norm is defined by*

$$\|x\|_{[U(\cdot) \to V]} = \left\| t \mapsto \|x(t, \cdot)\|_{U(t)} \right\|_V .$$

The spaces $[U \to V](w)$ are important examples of spaces with mixed family-norms: If we define $U(t) := U(w(t, \cdot))$, we have $[U(\cdot) \to V] = [U \to V](w)$. Observe that in this example the family $U(\cdot)$ is T-measurable, if U is T-measurable!

The approach to T-measurable families by means of the Luxemburg-Gribanov theorem seems to succeed only in the trivial case of countable families:

Definition 4.3.3. *A family of sets $U(\cdot)$ is said to be* countable measurable *on T, if the mapping $t \mapsto U(t)$ is measurable and essentially countable-valued, i.e. if there exist a countable number of sets U_n and measurable pairwise disjoint $T_n \subseteq T$ with $\bigcup T_n = T$, such that $U(t) = U_n$ for almost all $t \in T_n$.*

Theorem 4.3.1. *Let $U(\cdot)$ be a family of pre-ideal spaces over S, countable measurable on T. Then for any measurable function x on $T \times S$ the function $t \mapsto \|x(t, \cdot)\|_{U'(t)}$ is measurable.*

Proof. Put $B(t) = \{|u| : u \in U(t),\ \|u\|_{U(t)} \leq 1\}$ in Theorem A.3.3. □

Corollary 4.3.1. *Let $U(\cdot)$ be a family of perfect ideal spaces over S, countable measurable on T. Then $U(\cdot)$ is T-measurable.*

Proof. $U(t) = U''(t)$ with the same norm by Corollary 3.4.6. Apply Theorem 4.3.1 for $U'(t)$. □

For practice, another approach to T-measurable families is more promising. At first, we can say analogously to Lemma 4.1.2:

Definition 4.3.4. *Let $U(t)$ for almost all $t \in T$ be a normed linear space of measurable functions over S. We will call the family $U(\cdot)$ T-*simple measurable, *if for any simple measurable function x, satisfying $x(t,\cdot) \in U(t)$ for almost all $t \in T$, we have that $t \mapsto \|x(t,\cdot)\|_{U(t)}$ is measurable.*

Lemma 4.3.1. *Let $U(t)$ for almost all $t \in T$ be an almost perfect pre-ideal space over S. Then $U(\cdot)$ is T-measurable, if and only if $U(\cdot)$ is T-simple measurable.*

Proof. Let x be measurable with $x(t,\cdot) \in U(t)$ for almost all t. Let x_n be the sequence of Lemma 4.1.1. By Corollary 3.2.5 we have $\|x_n(t,\cdot)\|_{U(t)} \to \|x(t,\cdot)\|_{U(t)}$ for almost all t. □

A property, which in many cases is almost trivial to check, is the following one. For example, any countable measurable family has this property, but it is not hard to find other natural examples:

Definition 4.3.5. *Let $U(t)$ for almost all $t \in T$ be a normed linear space of measurable functions over S. We will call the family $U(\cdot)$ T-*elementary measurable, *if for any simple measurable function x on S the function $y(t) = \|x\|_{U(t)}$ is measurable on each measurable subset of its set of definition.*

Each T-measurable family is T-elementary measurable, since $(t,s) \mapsto \chi_M(t)x(s)$ is product-measurable for any measurable set M. We will prove that for most families of regular pre-ideal spaces the converse is also true.

Definition 4.3.6. *A set $E \subseteq T \times S$ is called* elementary, *if it is equivalent to a set of the form $D = A \times B$, where $A \subseteq T$, $B \subseteq S$ are measurable, i.e. if $E \setminus D$ and $D \setminus E$ both are null sets. A set $E \subseteq T \times S$ is called* e-simple, *if it is the finite union of elementary sets. A simple function x of the form $x(t,s) = \sum_{k=1}^{n} a_k \chi_{E_k}(t,s)$ with elementary sets E_k is called* elementary.

The following observation is almost obvious. However, its proof is quite technical:

Lemma 4.3.2. *A function $x : T \times S \to Y$ is elementary, if and only if it satisfies a.e.*

$$x(t,s) = \sum_{k=1}^{n} \chi_{T_k}(t) x_k(s), \tag{4.2}$$

where $T_k \subseteq T$ are measurable and pairwise disjoint, and $x_k : S \to Y$ are simple.

Proof. If x has the form (4.2), then

$$x(t,s) = \sum_{k=1}^{n} \chi_{T_k}(t) \sum_{j=1}^{n_k} a_{jk} \chi_{S_{jk}}(s) = \sum_{j,k} a_{jk} \chi_{T_k \times S_{jk}}(t,s),$$

i.e. x is elementary. Conversely, let x be elementary, i.e. x satisfies (without loss of generality) for all t, s:

$$x(t,s) = \sum_{j=1}^{p} a_j \chi_{D_j \times E_j}(t,s) = \sum_{j=1}^{p} a_j \chi_{D_j}(t) \chi_{E_j}(s),$$

where $D_k \subseteq T$, $E_k \subseteq S$ are measurable. Let $n = 2^p$, and let M be a one-to-one mapping of $\{1, \ldots, n\}$ onto the set of all subsets of $\{1, \ldots, p\}$. Let

$$T_k = \Big(\bigcup_{j \in M(k)} D_j \Big) \setminus \Big(\bigcup_{j \notin M(k)} D_j \Big).$$

Then T_k are measurable, pairwise disjoint, and $\bigcup T_k = \bigcup D_j$, since to each $t \in \bigcup D_j$ there exists exactly one set $M(k)$ of indices j, such that $t \in D_j$, if and only if $j \in M(k)$. This also shows that for any $t \in T_k$

$$x(t,s) = \sum_{j \in M(k)} a_j \chi_{E_j}(s) =: x_k(s).$$

Since the T_k are pairwise disjoint this implies (4.2). □

Lemma 4.3.3. *If x is a measurable bounded function on $T \times S$, then there exists a sequence of elementary functions x_n converging to x a.e.*

Proof. We may assume without loss of generality that $T \times S$ is of finite measure (Otherwise we just replace the measure by a finite measure with the same measurable sets and the same null sets as before, which is possible by Corollary 2.2.6). In particular, this means that $x \in L_1(T \times S)$.
The measure on $T \times S$ is generated by the algebra of finite unions of elementary sets. Thus, by [15, Lemma III.8.3], the integrable elementary functions are dense in $L_1(T \times S)$. This implies that there exists a sequence x_n of elementary functions converging to x in $L_1(T \times S)$. By Corollary 3.2.1 a subsequence converges a.e. to x. □

In the proof of [15, Lemma III.11.16(b)] it is used that the sequence of Lemma 4.3.3 additionally satisfies a.e. the estimate $|x_n| \le |x|$. In particular, this would imply for $x = \chi_E$ that to any product-measurable set E of positive measure there exists an elementary set D of positive measure with $\mathrm{mes}(D \setminus E) = 0$. Sadly, this is false!

One counterexample was posted on my question to news by Robert Israel 1995. It seems that his example is a modification of Elina Ermolova's idea for a counterexample that I posted in the question (she was able to prove her example later by other means). Since Ermolova's counterexample is even more astonishing, we use Israel's ideas to prove it:

Example 4.3.1. Let $T = [0,1]$, $S = [-1,2]$. Let $C \subseteq [-1,1]$ be closed with empty interior, but of positive measure, and let

$$E = \{(t, t+c) : t \in T,\ c \in C\} = \{(t,s) \in T \times S : s - t \in C\}.$$

Obviously, E is measurable (a rotation of the plane maps measurable sets into measurable sets), hence by the Cavalieri principle $\mathrm{mes}E = (\mathrm{mes}T)(\mathrm{mes}C) > 0$.
Let $D = A \times B \subseteq T \times S$ be an elementary set such that $N = D \setminus E$ is a null set. We prove that this implies $\mathrm{mes}D = 0$.
Indeed, for $C^c = \mathbb{R} \setminus C$ we have

$$N = \{(t,s) : t \in A,\ s \in B,\ s - t \in C^c\},$$

hence by Fubini-Tonelli

$$\begin{aligned} 0 &= \mathrm{mes}N = \int_A \int_{-\infty}^{\infty} \chi_B(s)\chi_{C^c}(s-t)\,ds\,dt \\ &= \int_A \int_{-\infty}^{\infty} \chi_B(s+t)\chi_{C^c}(s)\,ds\,dt = \int_{C^c} \int_A \chi_B(s+t)\,dt\,ds. \end{aligned}$$

Thus, the function $x(s) = \int_A \chi_B(t+s)dt$ vanishes for almost all $s \in C^c$. x is continuous, since

$$|x(s) - x(\sigma)| \le \int_{-\infty}^{\infty} |\chi_B(t+s) - \chi_B(t+\sigma)|\,dt \to 0 \text{ for } \sigma \to s,$$

see e.g. [1, Lemma 2.5]. Since C^c is open and dense in $\mathbb{R}$, we thus have $x = 0$. But then Fubini-Tonelli implies

$$\begin{aligned} 0 &= \int_{-\infty}^{\infty} |x(s)|\,ds = \int_{-\infty}^{\infty} \int_A \chi_B(t+s)\,dt\,ds = \int_{-\infty}^{\infty} \int_{-\infty}^{\infty} \chi_A(t)\chi_B(t+s)\,dt\,ds \\ &= \int_{-\infty}^{\infty} \int_{-\infty}^{\infty} \chi_B(t+s)\,ds\,\chi_A(t)\,dt = \int_A \mathrm{mes}B\,dt = (\mathrm{mes}A)(\mathrm{mes}B), \end{aligned}$$

i.e. $D = A \times B$ is a null set.

We remark that the example of course must also hold for $T = S = [0,1]$, if we replace E by $E \cap (T \times S)$ (but it is not that easy to see that $E \cap (T \times S)$ must have positive measure for arbitrary C). For C being the modified Cantor set this is Ermolova's counterexample.

Definition 4.3.7. *Let $U(t)$ for almost all $t \in T$ be a pre-ideal space over S. We say that the family $U(\cdot)$ has* uniformly-simple regular support, *if there exists a sequence $S_n \subseteq S$ such that χ_{S_n} belongs to the regular part of the real form of $U(t)$ and $\bigcup S_n = \operatorname{supp} U(t)$ for almost all $t \in T$.*

Theorem 4.3.2. *Let $U(t)$ for almost all $t \in T$ be an almost perfect pre-ideal space. Let the family $U(\cdot)$ have uniformly-simple regular support. Then the family $U(\cdot)$ is T-measurable, if and only if it is T-elementary measurable.*

Proof. By Lemma 4.3.1 it suffices to show that U is T-simple measurable. Thus let a simple function x be given, which satisfies $x(t,\cdot) \in U(t)$ for almost all t. We have to show that $t \mapsto \|x(t,\cdot)\|_{U(t)}$ is measurable. Since we may replace $x(t,s)$ by $c\,|x(t,s)|$ with $c > 0$, we may assume without loss of generality that $0 \le x(t,s) \le 1$.
Since $U(\cdot)$ has uniformly-simple support, there exists a sequence $S_1 \subseteq S_2 \subseteq \dots \subseteq S$ such that χ_{S_n} belongs to the regular part of $U(t)$, and $\bigcup S_n = \operatorname{supp} U(t)$ for almost all $t \in T$. Consider the sequence $y_n(t,s) = \chi_{S_n}(s)x(t,s)$. Then for almost all t we have that $y_n(t,\cdot)$ converges a.e. monotonically increasing to $x(t,\cdot)$. Hence Corollary 3.2.5 implies $\|y_n(t,\cdot)\|_{U(t)} \to \|x(t,\cdot)\|_{U(t)}$ for almost all $t \in T$. Thus it suffices to show that each $t \mapsto \|y_n(t,\cdot)\|_{U(t)}$ is measurable. In other words: We may assume without loss of generality that additionally x vanishes outside $T \times S_n$ for some n.
By Lemma 4.3.3 we may approximate x a.e. by a sequence of elementary functions z_k. Then also $x_k(t,s) = \min\{1, \chi_{S_n}(s)z_k(t,s)\}$ converges a.e. to x. Since $|x_k(t,\cdot)| \le |\chi_{S_n}|$, Theorem 3.3.5 implies $\|x_k(t,\cdot)\|_{U(t)} \to \|x(t,\cdot)\|_{U(t)}$ for almost all $t \in T$. Thus we may assume additionally without loss of generality that x is elementary.
By Lemma 4.3.2 this means that x has the form

$$x(t,s) = \sum_{k=1}^{n} \chi_{T_k}(t)x_k(s)$$

with pairwise disjoint measurable sets $T_k \subseteq T$ and simple functions $x_k \ge 0$. Define

$$y_k(t) = \begin{cases} \|x_k\|_{U(t)} & \text{if } t \in T_k \\ 0 & \text{otherwise.} \end{cases}$$

Since $|\chi_{T_k}(t)x_k(s)| \le |x(t,s)|$, y_k is defined a.e. Since $U(\cdot)$ is T-simple measurable, each y_k is measurable. Furthermore, we have, since the T_k are pairwise disjoint,

$$y(t) = \|x(t,\cdot)\|_{U(t)} = \sum_{k=1}^{n} y_k(t).$$

Thus y is measurable. □

If we really can choose $|x_n| \leq |x|$ in Lemma 4.3.3 (which in general is not possible, recall Example 4.3.1), it is easy to see that we also can satisfy $|x_1| \leq |x_2| \leq \ldots \leq |x|$ (cf. proof of Lemma 4.1.1). From this, a simple modification of the proof shows that in Theorem 4.3.2 we can drop the assumption that $U(\cdot)$ has uniformly-simple regular support. Thus it seems that this condition is just needed for technical reasons.

Usually, the easiest way to check that a family has uniformly-simple regular support is to check that it has simple-regular support (Definition 4.1.3):

Proposition 4.3.1. *Let $U(t)$ for almost all $t \in T$ be a pre-ideal space over S with simple-regular support. If there exists a sequence $S_n \subseteq S$ of sets of finite measure with $\chi_{S_n} \in U(t)$ and $\bigcup S_n = \mathrm{supp} U(t)$ for almost all $t \in T$, then $U(\cdot)$ has uniformly-simple regular support.*

In particular, in view of Corollary 2.2.7, any constant family of spaces with simple-regular support has uniformly-simple regular support. Thus, Theorem 4.3.2 implies Theorem 4.1.2.
Another consequence is by Theorem 4.1.3:

Corollary 4.3.2. *Let $U(\cdot)$ be a T-elementary measurable family of (almost perfect ideal) Orlicz spaces, each of which is generated by a finite Young function (and equipped with either the Luxemburg or Orlicz norm independently). Then $U(\cdot)$ is T-measurable.*

Note, however, that the converse of Proposition 4.3.1 does not hold: There exists a pre-ideal space U without simple-regular support, such that $U(t) \equiv U$ has uniformly-simple regular support (in particular, Theorem 4.1.2 may be weaker than Theorem 4.3.2, even for constant families):

Example 4.3.2. Let $S = [0,1]$, and U consist of all measurable functions with finite norm

$$\|x\| = \int_S |x(s)|\, ds + \operatorname*{ess\,lim\,sup}_{s \to 0+} |x(s)|\,.$$

U has no simple-regular support, since the simple function $x \equiv 1$ satisfies $\|P_{[0,\delta]}x\| \geq 1$ for any $\delta > 0$, but $U(t) \equiv U$ has uniformly-simple regular support with $S_n = \{0\} \cup [n^{-1}, 1]$.

One might guess that analogous Theorem 4.1.5 the support of a space $[U(\cdot) \to V]$ with mixed family-norm is $\{(t,s) : t \in \mathrm{supp} V,\ s \in \mathrm{supp} U(t)\}$ for a proper representation of $\mathrm{supp} U(t)$. Surprisingly, this can not be proved, even if all spaces are ideal spaces:

Example 4.3.3. Let $T = S = [0,1]$. Sierpinski showed [42] (see also [40, 8.9(c)]) that by the continuum hypothesis one can construct a subset Q of $T \times S$, such that each $Q(t) = \{s : (t,s) \in Q\}$ contains only countable many points, but such that every set $\{t : (t,s) \in Q\}$ contains almost all points of T. Now let $X = [U(\cdot) \to V]$ with $V = L_\infty(T)$, where $U(t)$ is the restriction of $L_\infty(S)$ to functions satisfying

$$\operatorname{ess} \limsup_{s \to s_0} |x(s)| = 0 \qquad (s_0 \in Q(t)).$$

Obviously, $U(t)$ is an ideal space. We state that $X = \{0\}$:
Indeed, let $x \in X$ be arbitrary. Let $s_0 \in [0,1)$, and $s_n \to s_0^+$. The sequence

$$y_n(t) = \frac{1}{s_n - s_0} \int_{s_0}^{s_n} |x(t,s)|\, ds$$

is bounded by $\|x\|_X$ and for all t with $(t, s_0) \in Q$ (i.e. for almost all t), we have $y_n(t) \to 0$ by the definition of $U(t)$. Hence, if we put

$$z(s) = \int_T |x(t,s)|\, dt,$$

Fubini-Tonelli and Lebesgue's dominated convergence theorem implies

$$\frac{1}{s_n - s_0} \int_{s_0}^{s_n} z(s) ds = \int_T y_n(t) dt \to 0.$$

But this implies that $z(s_0) = 0$, if s_0 is a Lebesgue point of z (see e.g. [40, Theorem 7.10]). Since almost all $s_0 \in [0,1)$ are Lebesgue points [40, Theorem 7.6], we have $z = 0$ a.e., whence $x = 0$ a.e.
Finally, we show that $\operatorname{supp} U(t) = S$ for each t: Indeed, $Q(t)$ is just countable, say $Q(t) = \{q_n : n \in \mathbb{N}\}$. Given $\varepsilon > 0$, let $C = \bigcup_n (q_n - \varepsilon 2^{-n}, q_n + \varepsilon 2^{-n})$ and $x = \chi_{S \setminus C}$. Then $x \in U(t)$. Since $\operatorname{mes} C \le \sum 2\varepsilon 2^{-n} = 2\varepsilon$, we see that $\operatorname{mes}(S \setminus \operatorname{supp} U(t)) < 2\varepsilon$, or $\operatorname{supp} U(t) = S$ as stated.

For the support of a space with mixed family-norm we thus state only a trivial inclusion condition:

Lemma 4.3.4. *If $x \in [U(\cdot) \to V]$, then for almost all t we have*

1. *$x(t,s) = 0$ for almost all s, if $t \notin V$, and*
2. *$x(t,s) = 0$ for almost all $s \notin \operatorname{supp} U(t)$.*

Proof. For almost all t we have that $x(t,\cdot) \in U(t)$, whence the second statement is immediate. Furthermore, $y(t) = \|x(t,\cdot)\|_{U(t)}$ belongs to V, thus for almost all $t \notin \operatorname{supp} V$ we have $y(t) = 0$, whence $x(t,\cdot) = 0$ a.e. □

Only for countable measurable families we sharpen the result:

Theorem 4.3.3. *Let $X = [U(\cdot) \to V]$ be defined, such that $U(\cdot)$ is a countable measurable family of pre-ideal spaces. Then*

$$\mathrm{supp} X = \{(t,s) : t \in \mathrm{supp} V,\ s \in \mathrm{supp} U(t)\},$$

where for $U(t_1) = U(t_2)$ also $\mathrm{supp} U(t_1) = \mathrm{supp} U(t_2)$.

Proof. There exist countable many U_n and measurable pairwise disjoint sets T_n with $T = \bigcup T_n$, such that $U(t) \equiv U_n$ for almost all $t \in T_n$. Let $S_n = \mathrm{supp} U_n$. We have to show that $\mathrm{supp} X = M$, where

$$M := \bigcup_n (T_n \cap \mathrm{supp} V) \times S_n.$$

If $x \in X$, Lemma 4.3.4 implies that $y = \chi_{(T \times S) \setminus M} x$ for almost all t satisfies $y(t,s) = 0$ for almost all s. Thus $y = 0$, and $\mathrm{supp} X \subseteq M$. For the converse, let $E \subseteq M$ have positive measure. Then there is some n such that $E_n = E \cap ((T_n \cap \mathrm{supp} V) \times S_n)$ has positive measure. Consider on $T_n \times S_n$ the space $[U_n \to V]$. By Theorem 4.1.5 it contains a function x, such that $E_n \cap \mathrm{supp} x$ has positive measure. The trivial extension of x to $T \times S$ belongs to X, and $E \cap \mathrm{supp} x \supseteq E_n \cap \mathrm{supp} x$ has positive measure. Thus $\mathrm{supp} X \supseteq M$. □

Since any semi-perfect pre-ideal space is almost α-perfect for some $\alpha > 0$ by Theorem 3.2.2, Theorem 4.1.6 is a special case of the following

Theorem 4.3.4. *Let $U(\cdot)$ be T-measurable, such that almost all $U(t)$ are α-perfect pre-ideal spaces (with the same α), and let V be a semi-perfect pre-ideal space over T. Then $[U(\cdot) \to V]$ is a semi-perfect ideal space.*

Proof. By Corollary 3.2.4 it suffices to show that $X = [U(\cdot) \to V]$ is semi-perfect. We use Lemma 3.2.7. Thus we assume that X is real-valued and that $0 \le x_1 \le x_2 \le \ldots$ is a sequence in X with bounded norms, and $x = \sup x_n$. Then $y_n(t) = \|x_n(t,\cdot)\|_{U(t)}$ satisfies $0 \le y_1 \le y_2 \le \ldots$ a.e., and $\|y_n\|_V$ is bounded. Since V is semi-perfect, we have $y = \sup y_n \in V$ by Lemma 3.2.7. In particular, for almost all $t \in T$ we have

$$\sup_n \|x_n(t,\cdot)\|_{U(t)} \le y(t) < \infty$$

and $0 \le x_1(t,\cdot) \le x_2(t,\cdot) \le \ldots$ a.e. Since $U(t)$ is α-perfect, this implies $x(t,\cdot) = \sup x_n(t,\cdot) \in U(t)$ and $\|x(t,\cdot)\|_{U(t)} \le \alpha y(t)$ by Lemma 3.2.7. But now $\alpha y \in V$ implies $x \in X$. □

Theorem 4.3.5. *Let $U(\cdot)$ be T-measurable, such that almost all $U(t)$ are almost α-perfect pre-ideal spaces, and let V be an almost β-perfect pre-ideal space over T. Then $[U(\cdot) \to V]$ is an almost $\alpha\beta$-perfect pre-ideal space.*

Proof. We use Lemma 3.2.7. To this end, assume that $X = [U(\cdot) \to V]$ is real-valued and that $0 \le x_1 \le x_2 \le \dots$ with $x = \sup x_n \in X$. For almost all t we have that $x_n(t,\cdot)$ converges a.e. monotonically increasing to $x(t,\cdot)$ and that $U(t)$ is almost α-perfect. Thus, Corollary 3.2.5 implies, if we put $y_n(t) = \|x_n(t,\cdot)\|_{U(t)}$ and $y(t) = \|x(t,\cdot)\|_{U(t)}$, that for almost all t

$$\alpha^{-1}y(t) \le \lim_{n\to\infty} y_n(t) = z(t) \le y(t). \tag{4.3}$$

Since y_n converges a.e. monotonically increasing to z, and $z \in V$ by $z \le y \in V$, and V is almost β-perfect, we have by Lemma 3.2.7 that

$$\|z\|_V \le \beta \liminf_{n\to\infty} \|y_n\|_V ,$$

which implies in view of (4.3)

$$\|x\|_X = \|y\|_V \le \|\alpha z\|_V \le \alpha\beta \liminf_{n\to\infty} \|y_n\|_V = \alpha\beta \liminf_{n\to\infty} \|x_n\|_X .$$

Thus, Lemma 3.2.7 implies the statement. □

Next, we want to prove that $[U(\cdot) \to V]$ is regular, if all $U(t)$ and V are regular.

Lemma 4.3.5. *If $D_n \subseteq T \times S$ are product-measurable, and*

$$D_n^t = \{s \mid (t,s) \in D_n\}, \qquad z_n(t) = \text{mes} D_n^t,$$

then $\text{mes} D_n \to 0$ *implies* $z_n \to 0$ *in measure.*

Proof. By Fubini-Tonelli

$$\text{mes} D_n = \int_T \int_S \chi_{D_n}(t,s)\,ds\,dt = \int_T |z_n(t)|\,dt.$$

□

Theorem 4.1.8 is a consequence of the following

Theorem 4.3.6. *Let $U(\cdot)$ be a T-measurable family, such that almost all $U(t)$ are regular pre-ideal spaces. Let V be a regular pre-ideal space over T. Then $[U(\cdot) \to V]$ is a regular pre-ideal space.*

Proof. If X is not inner-regular, there exist $x \in X$, $\varepsilon > 0$ and measurable $D_n \subseteq T \times S$ with $\text{mes} D_n \to 0$, such that

$$\left\| t \mapsto y_n(t) = \|\chi_{D_n}(t,\cdot)x(t,\cdot)\|_{U(t)} \right\|_V \ge \varepsilon \text{ for all } n \in \mathbb{N}.$$

Since $y_n(t) \le \|x(t,\cdot)\|_{U(t)} = y(t)$, and y lies in the regular part of V, by Theorem 3.3.5 no subsequence of y_n may tend to zero a.e.

But on the other hand by Lemma 4.3.5 there exists a subsequence, such that for almost all t we have $\text{mes} D_{n_k}^t \to 0$, where $D_n^t = \{s \mid (t,s) \in D_n\}$. This implies that $u_k^t(s) = \chi_{D_{n_k}}(t,s)x(t,s)$ tends to zero in measure for almost all t. Since $|u_k^t| \le |x(t,\cdot)|$ and $x(t,\cdot)$ lies in the regular part of $U(t)$ for almost all t, by Theorem 3.3.5 we have $\|u_k^t\|_{U(t)} \to 0$ for almost all t. But this means $y_{n_k} \to 0$ almost everywhere, a contradiction. Thus X is inner-regular.
Since T, S are σ-finite, there exist increasing sequences $S_1 \subseteq S_2 \subseteq \ldots$, $T_1 \subseteq T_2 \subseteq \ldots$ of sets of finite measure with $S = \bigcup S_n$, $T = \bigcup T_n$. To prove that X is outer-regular, it suffices to show that for any given $x \in X$ we have

$$\lim_{n\to\infty} \|P_{(T\times S)\setminus(T_n\times S_n)}x\|_X = 0.$$

Since

$$|P_{(T\times S)\setminus(T_n\times S_n)}x| \le |P_{T\times(S\setminus S_n)}x| + |P_{(T\setminus T_n)\times S}x|,$$

it suffices to show that

$$\lim_{n\to\infty} \|P_{T\times(S\setminus S_n)}x\|_X = \lim_{n\to\infty} \|P_{(T\setminus T_n)\times S}x\|_X = 0.$$

For the first limit we use that by Lemma 3.3.1

$$y_n(t) = \|P_{S\setminus S_n}x(t,\cdot)\|_{U(t)} \to 0 \text{ for almost all } t,$$

hence, $|y_n(t)| \le y(t) = \|x(t,\cdot)\|_{U(t)}$ implies by Theorem 3.3.5

$$\|P_{T\times(S\setminus S_n)}x\|_X = \|y_n\|_V \to 0.$$

For the second limit we observe that

$$\|P_{(T\setminus T_n)\times S}x\|_X = \|P_{T\setminus T_n}y\|_V,$$

so we just have to apply Lemma 3.3.1. □

Corollary 4.3.3. *Let $U(\cdot)$ be a T-measurable family of pre-ideal spaces, and V be a pre-ideal space. Let x be product-measurable. If $x(t,\cdot)$ belongs to the regular part of $U(t)$ for almost all t, and $y(t) = \|x(t,\cdot)\|_{U(t)}$ to the regular part of V, then x belongs to the regular part of $[U(\cdot) \to V]$.*

Proof. Denote the regular part of $U(t)$ and V by $U_0^0(t)$ and V_0^0 respectively. Theorem 4.3.6 implies that $Y = [U_0^0(\cdot) \to V_0^0]$ is regular, hence contained in the regular part of $[U(\cdot) \to V]$. The conditions on x ensure that $x \in Y$. □

Now we want to calculate the associate space to a space with mixed family-norm. In view of Theorem 4.1.10 we define:

Definition 4.3.8. *If $X = [U(\cdot) \to V]$ is defined with pre-ideal spaces $U(t)$ and V, then X is called* natural associate, *if $Y = [U'(\cdot) \to V']$ is defined and $X' = Y$ with the same norm.*

One direction is almost trivially satisfied without any conditions:

Lemma 4.3.6. *If $X = [U(\cdot) \to V]$ and $Y = [U'(\cdot) \to V']$ are defined, then $\|y\|_{X'} \le \|y\|_Y$ for any $y \in Y$.*

Proof. Denote the underlying measure spaces by T resp. S. For any $x \in X$ we have by Fubini-Tonelli

$$\int_{T\times S} |y(t,s)|\,|x(t,s)|\,d(t,s) \le \int_T \|y(t,\cdot)\|_{U'(t)}\,\|x(t,\cdot)\|_{U(t)}\,dt \le \|y\|_Y\,\|x\|_X\,,$$

which implies the statement. □

In Lemma 4.3.6 the T-measurability of $U'(\cdot)$ is needed to write down the second integral.

We emphasize that surprisingly Lemma 4.3.6 does not imply $Y \subseteq X'$, since it does not ensure that y must vanish a.e. outside supp$X$, i.e. it may happen that supp$X \not\subseteq$ supp$Y$:

Example 4.3.4. Define $X = [U(\cdot) \to V]$ with $T = S = [0,1]$ as in Example 4.3.3. In particular, $X = \{0\}$, but supp$V = T$ and supp$U(t) = S$ for all t. Since $U(t)$ is a subspace of $L_\infty(S)$ with full support, we have at least that $L_1(S)$ is continuously embedded in $U'(t)$. Since $V' = L_1(T)$, we have that $L_1(T \times S) = [L_1(S) \to V']$ is continuously embedded in $Y = [U'(\cdot) \to V']$. In particular, supp$Y = T \times S$.

The example also shows that not all spaces with mixed family-norm need to be natural associate.

For $T = \mathbb{N}$ we prove the converse of Lemma 4.3.6. More interesting than the statement itself is the idea of the proof, which we will generalize (be aware that for uncountable T the proof makes also use of the axiom of choice):

Theorem 4.3.7. *Assume that any real function x is measurable on $T \times S$, provided that $x(t,\cdot)$ is measurable on S for almost all t. Let $U(t)$ for almost all $t \in T$ be a pre-ideal space over S, and V be a pre-ideal space over T. Then the associate space of $X = [U(\cdot) \to V]$ is continuously embedded in $Y = [U'(\cdot) \to V']$ with embedding constant 1.*

Proof. First, we remark that X and Y are defined, since any real function on T must be measurable (unless S is trivial, which we may exclude). Now let $y \in X'$, $\|y\|_{X'} \le 1$. We have to prove that this implies $y \in Y$, $\|y\|_Y \le 1$. Observe that by Lemma 4.3.4 we have supp$y(t,\cdot) \subseteq$ supp$U(t)$ for almost all t, and the support of $z(t) = \|y(t,\cdot)\|_{U'(t)} = 0$ is a subset of supp$V$. Hence, if the statement is false, there is some $v \in V$, $\|v\|_V \le 1$ with

$$\int_T \|y(t,\cdot)\|_{U'(t)} |v(t)|\, dt > 1.$$

Since

$$\|y(t,\cdot)\|_{U'(t)} = \sup_{\|u\|_{U(t)} \le 1} \int_S |y(t,s)|\, |u(s)|\, ds,$$

we can find for any t a function $u_t \in U(t)$, $\|u_t\|_{U(t)} \le 1$ such that we still have

$$\int_T \left(\int_S |y(t,s)|\, |u_t(s)|\, ds \right) |v(t)|\, dt > 1 \tag{4.4}$$

(Lemma A.2.2). Now define $x(t,s) = |u_t(s)|\,|v(t)|$ (if T is uncountable, the axiom of choice is needed to ensure that x is a function). By assumption, x is measurable. But then $x \in X$ with

$$\|x\|_X = \|t \mapsto |v(t)| \ \|u_t\|_{U(t)}\|_V \le \|v\|_V \le 1.$$

Hence $\|y\|_{X'} \le 1$ implies

$$\int_{T\times S} |x(t,s)|\, |y(t,s)|\, d(t,s) \le 1,$$

which is a contradiction to (4.4). □

The proof of Theorem 4.3.7 fails without the assumption on $T \times S$, since the function x 'constructed' there need not be product-measurable.
It turns out that the problem of constructing such a measurable function x is similar to T-measurability. Thus we try to apply the Luxemburg-Gribanov theorem. In this way we get (even without the axiom of choice) the following generalization of Theorem 4.1.10:

Theorem 4.3.8. *Let $U(\cdot)$ be a family of pre-ideal spaces over S, which is countable measurable on T. Then, if $X = [U(\cdot) \to V]$ is defined, it is natural associate, i.e. $Y = [U'(\cdot) \to V']$ is defined and coincides with X' (with the same norm).*

Proof. Y is defined by Theorem 4.3.1. The difficulty described in Example 4.3.4 can not occur, since Theorems 4.3.3 and 3.4.6 imply $\mathrm{supp} X' = \mathrm{supp} X = \mathrm{supp} Y$. Hence the embedding $X' \subseteq Y$ with embedding constant 1 is an immediate consequence of Lemma 4.3.6.
Thus, we just have to find a way to save the statement of Theorem 4.3.7. For this ending, let $y \in X'$, $\|y\|_{X'} \le 1$. We have to prove that this implies $y \in Y$, $\|y\|_Y \le 1$. Since we already know that $\mathrm{supp} X' = \mathrm{supp} Y$, we otherwise find some $v \in V$, $\|v\|_V \le 1$ with

$$\int_T \|y(t,\cdot)\|_{U'(t)} |v(t)|\, dt > 1.$$

By Lemma A.2.2, we have

$$\int_T z(t)\,|v(t)|\,dt > 1$$

for some measurable $0 \leq z(t) \leq \|y(t,\cdot)\|_{U'(t)}$, such that $h(t) = \|y(t,\cdot)\|_{U'(t)} - z(t) > 0$, whenever $y(t,\cdot) \neq 0$. Now we apply Theorem A.3.4 for

$$B(t) = \{|u| : u \in U(t),\ \|u\|_{U(t)} \leq 1\},$$

and find, that there exists a nonnegative measurable function u on $T \times S$, such that

$$\int_S |y(t,s)|\,u(t,s)ds \geq \|y(t,\cdot)\|_{U'(t)} - h(t) = z(t)$$

and $u(t,\cdot) \in B(t)$ for almost all $t \in T$. Then $x(t,s) = u(t,s)\,|v(t)|$ belongs to the real form of X, and

$$\|x\|_X = \|t \mapsto |v(t)|\ \ \|u(t,\cdot)\|_{U'(t)}\|_V \leq \|v\|_V = 1.$$

Hence $\|y\|_{X'} \leq 1$ implies

$$\int_{T\times S} |x(t,s)|\,|y(t,s)|\,d(t,s) \leq 1.$$

But on the other hand we have

$$\int_T \left(\int_S |y(t,s)|\,u(t,s)ds\right)|v(t)|\,dt \geq \int_T z(t)\,|v(t)|\,dt > 1,$$

which is a contradiction by Fubini-Tonelli. □

4.4 Calculus with Ideal-valued Functions

We are now considering measurable functions x, which take values in an ideal space X. We will write $x(t)(s) = x(t,s)$. Here we have the problem, that $x(t)$ is only defined up to a set of measure zero. In [8, Theorem 2.1(ii)] it is stated that for the Lebesgue measure and $X = L_p$ each function $x(\cdot,\cdot)$ with the property that the function $t \mapsto x(t,\cdot)$ is continuous as a function from some compact interval into X is already product-measurable. This is false, as can be seen by a counterexample of Sierpinski [42] (see also [40, 8.9(c)]):
By the continuum hypothesis one can construct a subset Q of $T \times S = [0,1] \times [0,1]$ such that each $Q_t = \{s : (t,s) \in Q\}$ contains only countable many points, but that Q is not product-measurable. Obviously, $x(t)(s) = \chi_Q(t,s)$ is not product-measurable, but $x(t) = 0$ a.e.

However, the proof of [8] (see also [22, p.68-70]) shows for $X = L_p$, that there exists a suitable representation for each $x(t)$, such that $x(t,s) = x(t)(s)$ is product-measurable. We may generalize this for measurable functions and arbitrary pre-ideal spaces (even for vector-valued ones):

Since a pre-ideal space X need not be complete, we denote its completion by $\overline{X}$. A function $x : T \to X$ is then called measurable, of course, if it is measurable as a function $T \to \overline{X}$.

Theorem 4.4.1. *Let T, S be σ-finite measure spaces, and X be a pre-ideal space over S. Let $x : T \to X$ be measurable. Then there exists a $T \times S$-measurable function y with $x(t) = y(t,\cdot)$ for all $t \in T$.*

Proof. First, assume that $x = \sum y_k \chi_{E_k}$ is a simple function. Then the statement is true with $y(t,s) = \sum y_k(s)\chi_{E_k}(t)$. For the general case, let $T_1 \subseteq T$, $S_1 \subseteq S$ be arbitrary sets of finite measure. Since x is measurable, there exists a sequence w_n of simple functions $w_n : T \to \overline{X}$, such that $w_n \to x$ a.e. on T_1. Since X is dense in $\overline{X}$, there exist simple functions $x_n : T \to X$ with $|x_n - w_n| \leq n^{-1}$ on T. Then $x_n \to x$ a.e. on T_1. Let $y_n(t,s) = \chi_{S_1}(s)x_n(t)(s)$ be measurable on $T \times S$. Then for almost all $t \in T_1$ we have

$$\|P_{S_1}x(t) - y_n(t,\cdot)\| = \|P_{S_1}(x(t) - x_n(t))\| \leq \|x(t) - x_n(t)\| \to 0 \text{ for } n \to \infty,$$

in particular, $y_n(t,\cdot)$ is a Cauchy sequence in the restriction U of X to functions vanishing a.e. outside S_1. This implies that y_n is a Cauchy sequence in the (complete) metric linear space of measurable functions on $T_1 \times S_1$. Indeed, otherwise there exist $\delta, \varepsilon > 0$ and $m_k \geq n_k \to \infty$ with

$$\operatorname{mes}\{(t,s) \in T_1 \times S_1 : |y_{n_k}(t,s) - y_{m_k}(t,s)| \geq \delta\} \geq \varepsilon \qquad (k = 1,2,\ldots),$$

which is not possible, because $z_k = y_{n_k} - y_{m_k}$ tends to zero in measure on $T_1 \times S_1$ by Lemma 3.1.3. Thus, a subsequence of y_n converges almost everywhere in $T_1 \times S_1$ to some function, which is measurable on $T_1 \times S_1$: $y_{n_k} \to y^*$.

For almost all $t \in T_1$ we have $y_{n_k}(t,\cdot) \to y^*(t,\cdot)$ a.e., and $y_{n_k}(t,\cdot) \to P_{S_1}x(t)$ in U, hence $P_{S_1}x(t) = y^*(t,\cdot)$ a.e. on S_1 by Theorem 2.2.3. For those t define $y(t,s) = y^*(t,s)$ for $s \in S_1$, and for the remaining t let $y(t,s) = x(t)(s)$ with some arbitrary representation of $x(t)$. Then y is measurable on $T_1 \times S_1$, and $x(t) = y(t,\cdot)$ a.e. on S_1 for all $t \in T_1$. Hence, there exists a representation of $x(t)$, such that $y(t,s) = x(t)(s)$ is product-measurable on $T_1 \times S_1$. Since T and S are σ-finite, and T_1, S_1 were arbitrary, the extension to $T \times S$ is immediate. $\square$

Of course, in practice the problem occurs to check measurability of an abstract function. Especially, the question arises, whether the converse of Theorem 4.4.1 is true. For regular ideal spaces the answer is in the affirmative. For $L_p(S)$ the corresponding theorem is stated in [15, Lemma III.11.16(b)], but the proof there contains a flaw as mentioned in the remark following Lemma 4.3.3:

Theorem 4.4.2. *Let T, S be σ-finite measure spaces, and X be a regular pre-ideal space over S. Then, the abstract function $x : T \to X$ is measurable, if and only if there exists a function y, which is measurable on $T \times S$, such that $x(t) = y(t,\cdot)$.*

Proof. One direction follows by Theorem 4.4.1. For the other direction let a product-measurable function y be given such that $x(t) = y(t,\cdot) \in X$ for almost all $t \in T$. We have to prove that $x : T \to X$ is measurable.
By Corollary 2.2.7 there exists a sequence of measurable sets $S_1 \subseteq S_2 \subseteq \ldots \subseteq S$ with $\bigcup S_n = \operatorname{supp} X$, such that χ_{S_n} belongs to the real form of X.
Consider the sequence

$$y_n(t,s) = \begin{cases} y(t,s) & \text{if } |y(t,s)| \le n \text{ and } s \in S_n, \\ 0 & \text{otherwise.} \end{cases}$$

Let $x_n(t) = y_n(t,\cdot)$. Then for almost all $t \in T$ we have $x_n(t) \to x(t)$ a.e., and $|x_n(t)| \le |x(t)|$ a.e., hence Theorem 3.3.5 implies $x_n(t) \to x(t)$ in X. Thus, it suffices to prove that each $x_n : T \to X$ is measurable. In other words: We may assume additionally that there exists some n with $|y(t,s)| \le n$ and $y(t,s) = 0$ for $s \notin S_n$.
Now by Lemma 4.3.3 there exists a sequence of elementary functions z_k, which converges a.e. to y. Define

$$y_k(t,s) = \begin{cases} z_k(t,s) & s \in S_n \text{ and } |z_k(t,s)| \le n, \\ n\,|z_k(t,s)|^{-1}\, z_k(t,s) & s \in S_n \text{ and } |z_k(t,s)| > n, \\ 0 & s \notin S_n. \end{cases}$$

Then also y_k is elementary and converges a.e. to y. Let $x_k(t) = y_k(t,\cdot)$. Then for almost all $t \in T$ we have $x_k(t) \to x(t)$ a.e., and $|x_k(t)| \le n\chi_{S_n}$ a.e., whence $x_k(t) \to x(t)$ in X by Theorem 3.3.5. Thus, it suffices to prove that

$x_k(t)$ is measurable, i.e. we may assume additionally that y is elementary. By Lemma 4.3.2 this means that we may assume that y has the form

$$y(t,s) = \sum_{k=1}^{n} \chi_{T_k}(t) y_k(s),$$

where $T_k \subseteq T$ are measurable and pairwise disjoint, and y_k is simple. But since the T_k are pairwise disjoint, we have $x(t) = \sum \chi_{T_k}(t) y_k$, which shows that $x : T \to X$ is a simple function, whence measurable. □

Theorem 4.4.2 may fail even for perfect ideal spaces, if they are not regular:

Example 4.4.1. Let $T = S = [0,1]$, and $X = L_\infty(S)$. The function $y(t,s) = \chi_{[0,t]}(s)$ is product-measurable, but $x(t) = y(t,\cdot) = \chi_{[0,t]}$ is not measurable as a function $T \to X$, since it is not essentially separable valued: This follows by $\|x(t) - x(\tau)\| = 1$ for $t \neq \tau$ and Lemma A.1.1.

Theorem 4.4.3. *Let T, S be σ-finite measure-spaces, X be an ideal space over S, and $x : T \to X$ be integrable. If we choose a product-measurable representation of $x(t,s) = x(t)(s)$, we have that $x(\cdot, s)$ is integrable with*

$$\left(\int_T x(t)dt\right)(s) = \int_T x(t,s)dt \text{ for almost all } s \in S.$$

Proof. At first, we prove that the function $z : T \to X$, defined by $z(t) = |x(t,\cdot)|$ is also integrable. Indeed, z is measurable, since if x_n is a sequence of simple functions converging to x a.e., we have that $z_n(t) = |x_n(t,\cdot)|$ converges to z a.e. Since $t \mapsto \|z(t)\| = \|x(t)\|$ is integrable, also z must be integrable.
Let $y \in X'$, $y \geq 0$ be arbitrary. Define a continuous linear operator on X by

$$Ax = \int_S y(s)x(s)ds.$$

Then the following integrals must exist, be finite and equal:

$$\int_S y(s) \left[\int_T z(t)dt\right](s)ds = A\int_T z(t)dt = \int_T Az(t)dt = \int_T \int_S y(s)\,|x(t,s)|\,dsdt.$$

Since the last integral is finite, we have that $(t,s) \mapsto y(s)x(t,s)$ is integrable on $T \times S$ by the theorem of Fubini-Tonelli. On the one hand this shows by Fubini's theorem that $y(s)x(\cdot,s)$ is integrable for almost all $s \in S$, and thus that $x(\cdot,s)$ is integrable for almost all $s \in S$, if we choose y such that $\operatorname{supp} y = \operatorname{supp} X'$, which is possible by Theorem 3.4.6.
On the other hand we now are allowed to use Fubini's theorem to interchange the order of integration, and get analogously

$$\int_S y(s) \left[\int_T x(t)dt\right](s)ds = \int_T \int_S y(s)x(t,s)dsdt = \int_S \int_T y(s)x(t,s)dtds.$$

Hence

$$\int_S y(s)\left(\left[\int_T x(t)dt\right](s) - \int_T x(t,s)dt\right) ds = 0 \qquad (y \in X',\ y \geq 0).$$

The statement follows now by Theorem 3.4.10. □

As a sample application of the abstract theory, we give a short proof of Minkowski's integral inequality:

Corollary 4.4.1. *Let T, S be σ-finite measure spaces. If y is measurable on $T \times S$, and $1 \leq p < \infty$, then*

$$\left(\int_S \left(\int_T |y(t,s)|\, dt\right)^p ds\right)^{1/p} \leq \int_T \left(\int_S |y(t,s)|^p\, ds\right)^{1/p} dt. \tag{4.5}$$

Proof. Let $X = L_p(S)$. Define a function $x : T \to X$ by $x(t)(s) = |y(t,s)|$. By Theorem 4.4.2, x is measurable. Thus, if the right hand side of (4.5) is finite, we have that $t \mapsto \|x(t)\|$ is integrable, whence $x : T \to X$ is integrable. By Theorem 4.4.3 this implies

$$\left\|\int_T |y(t,\cdot)|\, dt\right\| = \left\|\int_T x(t)dt\right\| \leq \int_T \|x(t)\|\, dt = \int_T \|\, |y(t,\cdot)|\, \|\, dt,$$

where all expressions are defined and finite. This is the stated inequality. □

In the following, let I be some bounded or unbounded interval, and Y be some Banach space. We will say, that a function $x : I \to Y$ *satisfies FTC* (FTC=fundamental theorem of calculus), if x' exists a.e. on I and is locally integrable, such that for one (and thus all) $t_0 \in I$ we have

$$x(t) - x(t_0) = \int_{t_0}^t x'(\tau)d\tau \qquad (t \in I).$$

If Y is finite-dimensional, it is necessary and sufficient for x to satisfy FTC, that x is absolutely continuous. In the general case, this is not sufficient, since an absolutely continuous vector-valued function need not be differentiable anywhere, see [22, Remark before Theorem 3.8.6]. However, it is necessary that x is absolutely continuous, which follows by the following lemma (for the proof, see [22, Corollary 3.8.2]):

Lemma 4.4.1. *Let X be some Banach space, $x : \mathbb{R} \to X$ be integrable. Then for almost all $t \in \mathbb{R}$ we have*

$$\lim_{h \to 0} \frac{1}{h} \int_t^{t+h} \|x(\tau) - x(t)\|\, d\tau = 0.$$

Now, let X be an Y-valued ideal space over some σ-finite measure space S, and I be some interval. Then we define $B(I,X)$ as the space of all locally integrable functions $x : I \to X$, and $W(I,X)$ as the space of all functions $x : I \to X$, which satisfy FTC. Furthermore, we define $B_m(I,X)$ as the space of all measurable functions $x : I \times S \to Y$, such that $x(\cdot, s)$ is locally integrable for all s, and such that $t \mapsto x(t,\cdot)$ belongs to $B(I,X)$. Finally, let $W_m(I,X)$ be the space of all measurable functions $x : I \times S \to Y$, such that $x(\cdot,s)$ satisfies FTC for all s, where $(t,s) \mapsto \partial x(t,s)/\partial t$ belongs to $B_m(I,X)$.

Corollary 4.4.2. *A function $x : I \to X$ belongs to $B(I,X)$, if and only if it has some representation $x(t,s) = x(t)(s)$, which belongs to $B_m(I,X)$. In this case we have (a.e.)*

$$\int_{t_0}^{t} x(\tau)d\tau = \int_{t_0}^{t} x(\tau,\cdot)d\tau \qquad (t, t_0 \in I). \tag{4.6}$$

Proof. Let $x \in B(I,X)$. By Theorem 4.4.1 for $T = I$, x has some measurable representation. Since $I = \bigcup T_n$, where x is integrable on T_n, Theorem 4.4.3 implies, that $x(\cdot,s)$ is integrable on T_n for almost all s. Just put $x(t,s) = 0$ for the remaining s. Do this for all n. Then $x(\cdot,s)$ is integrable on T_n for all n and all s, i.e. $x(\cdot,s)$ is locally integrable for all s. By construction $x(\cdot,\cdot)$ is measurable, and $x(t) = x(t,\cdot)$ for all t. (4.6) holds by Theorem 4.4.3 for $T = [t_0, t]$ resp. $T = [t, t_0]$. $\square$

Remark 4.4.1. For fixed t_0 the function occuring in (4.6), $y(t,s) = \int_{t_0}^{t} x(\tau,s)d\tau$, is even measurable on $I \times S$. This follows by Fubini's theorem, since $y(t,s) = \int_I z(\tau,t,s)d\tau$, where $z(\tau,t,s) = w(\tau,t)x(\tau,s)$ with

$$w(\tau,t) = \begin{cases} 1 & t_0 \le \tau \le t, \\ -1 & t \le \tau < t_0, \\ 0 & \text{otherwise,} \end{cases}$$

is measurable on $I \times (I \times S)$.

Theorem 4.4.4. *A function $x : I \to X$ belongs to $W(I,X)$, if and only if it has some representation $x(t,s) = x(t)(s)$, which belongs to $W_m(I,X)$. In this case we have for almost all $t \in I$ that (a.e.)*

$$\frac{dx(t)}{dt} = \frac{\partial x(t,\cdot)}{\partial t}. \tag{4.7}$$

In particular, if we choose a measurable representation of the left hand side (which exists by Theorem 4.4.3), we have

$$\frac{dx(t)}{dt}(s) = \frac{\partial x(t,s)}{\partial t} \text{ for almost all } (t,s).$$

Proof. First, assume $x \in W(I,X)$. Since $x' = dx/dt \in B(I,X)$, it has some measurable representation $x'(t,s) = x'(t)(s)$, corresponding to Corollary 4.4.2. Fix some $t_0 \in I$. Then for all $t \in I$ we have a.e.

$$\int_{t_0}^{t} x'(\tau,\cdot)d\tau = \int_{t_0}^{t} x'(\tau)d\tau = x(t) - x(t_0).$$

Thus, if we define

$$x(t,s) = \int_{t_0}^{t} x'(\tau,s)d\tau + x(t_0)(s),$$

with some arbitrary representation of $x(t_0)$, we have $x \in W_m(I,X)$ (Remark 4.4.1) and $x(t,\cdot) = x(t)$ for all t.
Now, assume $x(t,s) = x(t)(s)$ lies in $W_m(I,X)$. At first, we show that $dx(t)/dt$ exists for almost all t. Indeed, by Corollary 4.4.2, we have that $y(t) = \partial x(t,\cdot)/\partial t$ belongs to $B(I,X)$ with

$$\|x(t+h) - x(t) - h\partial x(t,\cdot)/\partial t\| = \|\int_{t}^{t+h} \left[\frac{\partial x}{\partial t}(\tau,\cdot) - \frac{\partial x}{\partial t}(t,\cdot)\right] d\tau\|$$
$$= \|\int_{t}^{t+h} [y(\tau) - y(t)]\, d\tau\| \le |\int_{t}^{t+h} \|y(\tau) - y(t)\| d\tau|,$$

and by Lemma 4.4.1 for almost all t the right hand side is in $o(|h|)$. Thus for almost all t we have that $x'(t)$ exists, and (4.7) holds. Now by Corollary 4.4.2 we have

$$\int_{t_0}^{t} x'(\tau)d\tau = \int_{t_0}^{t} \frac{\partial x}{\partial t}(\tau,\cdot)d\tau = x(t,\cdot) - x(t_0,\cdot) = x(t) - x(t_0),$$

thus x satisfies FTC. □

We emphasize a special case: If Y is finite-dimensional, then for a measurable function $x : I \times S \to Y$ to be absolutely continuous w.r.t. the first variable, such that $t \mapsto \partial x(t,\cdot)/\partial t$ is locally integrable as a function from I into X, it is necessary and sufficient that $t \mapsto x(t,\cdot)$ satisfies FTC (as a function $I \to X$).

Similarly, we define $C(I,X)$ and $C^1(I,X)$ as the space of continuous resp. continuously differentiable functions $x : I \to X$. Let $C_m(I,X)$ be the space of measurable functions $x : I \times S \to Y$ such that $t \mapsto x(t,\cdot)$ belongs to $C(I,X)$, and $C^1_m(I,X)$ be the space of measurable functions $x : I \times S \to Y$, such that $x(\cdot,s)$ satisfies FTC for all s, where $(t,s) \mapsto \partial x(t,s)/\partial t$ belongs to $C_m(I,X)$. Then we have

Corollary 4.4.3. *With the identification $x(t,s) = x(t)(s)$ for some proper chosen representation we have that $C(I,X) = C_m(I,X)$ and $C^1(I,X) = C^1_m(I,X)$. Furthermore,* (4.6) *holds a.e. for functions in $C_m(I,X)$, and* (4.7) *holds a.e. for functions in $C^1_m(I,X)$ for any $t \in I$.*

Proof. To see that (4.7) even holds for all t, observe that it holds for almost all t, and both sides are continuous as functions from I into X. $\square$

5. Operators and Applications

5.1 Automatic Continuity of Linear Operators

As we mentioned in the beginning, without the axiom of choice it can not be shown that there exists an unbounded linear operator, which is defined on the whole of a Banach space. We want to discourage you using the axiom of choice by showing that for 'natural operators' defined on 'natural spaces' continuity is always automatic.
One result in this direction was Theorem 3.4.3, which shows in particular that integral functionals on ideal spaces are automatically bounded. Banach's theorem shows that for L_p-spaces the result also holds for integral operators. Its proof holds also for ideal spaces. Sadly, if the functions in consideration take values in infinite-dimensional spaces, we need an additional condition:

We consider the following situation: Let Z, W be Banach spaces, and S, T be measure spaces. Let X be a normed linear space of (classes of) measurable functions $S \to Z$. We say that the linear integral operator

$$Kx(t) = \int_S k(t,s)x(s)ds, \tag{5.1}$$

where the kernel k takes values in $\mathfrak{L}(Z, W)$, is *defined* on X, if for any $x \in X$ we have that $s \mapsto k(t,s)x(s)$ is integrable over S for almost all $t \in T$.
We say, two kernels k_1 and k_2 are *equivalent* on X, if for the corresponding integral operators K_1 resp. K_2 we have $K_1x = K_2x$ a.e. for any $x \in X$.
If (5.1) is defined on X, we say that it is *absolutely defined* on X, if some of its equivalent kernels k has the property that the integral operator

$$|K|\,x(t) = \int_S |k(t,s)|\,x(s)ds \tag{5.2}$$

is defined of X.

For finite-dimensional Z this is no additional condition:

Theorem 5.1.1. *If X is a finite-dimensional-valued pre-ideal space, then any integral operator defined on X is absolutely defined on X.*

Proof. We use the notation above. Choose a base $e_1, \ldots, e_N$ of Z. Since all norms on Z are equivalent, there exists a constants $C > 0$ such that for all $z \in Z$, $z = \sum z_n e_n$ with $|z| \leq 1$ we have $\max_n |z_n| \leq C$. In particular, for any $l \in \mathfrak{L}(Z, W)$ we have

$$|l| = \sup_{|z| \leq 1} |lz| \leq \sup_{|z_n| \leq C} |l(\sum_{n=1}^{N} z_n e_n)| \leq C \sum_{n=1}^{N} |l e_n|.$$

Now fix some $x \in X$. Since X is a pre-ideal space, also the functions $x_n(s) = |x(s)| e_n$ belong to X, whence $K x_n$ is defined a.e., hence

$$\infty > \sum_{n=1}^{N} \int_S |k(t,s) e_n| \, |x(s)| \, ds \geq \int_S C^{-1} |k(t,s)| \, |x(s)| \, ds$$

for almost all $t \in T$. □

It is not known to us, whether Theorem 5.1.1 holds also for any infinite-dimensional Z.

Now we may transfer Banach's theorem to ideal spaces:

Definition 5.1.1. *Let X be a linear subspace of some F-space B. Assume, the elements of X are (classes of) measurable functions with values in some Banach space. Then X is called* proper, *if $x_n \in X$, $x_n \to x$ in B, and $x_n \to y$ a.e. with $y \in X$ imply $x = y$.*

By Theorem 2.2.3 any complete subspace of an ideal space is proper, but obviously there are much more examples: Any 'natural' function space is proper.

An F-space of measurable functions is proper, if it has the W-property. The converse holds by the closed graph theorem for F-spaces, if the underlying measure space is finite (cf. remark before Lemma 3.1.2). But we emphasize that the W-property is not necessary, if the underlying measure space is infinite, not even for ideal spaces (see Example 3.1.3).
Thus, any ideal* space over a finite measure space without the W-property (e.g. the space of Example 3.1.2) can not be proper.

However, we should remark that there exist pre-ideal spaces, which are not proper (and contrary to Example 3.1.2, the axiom of choice is not needed to construct them): Consider the space X of Example 3.2.1 and the Cauchy-sequence x_n given there. Then x_n converges a.e. to the constant function 1, which belongs to X. However, x_n does not converge in norm to this function. But, since x_n is Cauchy, it thus must converge to a different element in B.

The following proof is mainly the same as that of Theorem 5.1.3:

Theorem 5.1.2. *Let X be an ideal space. Assume, a linear integral operator K is absolutely defined on X. If $K : X \to Y$, Y being some proper space, then K is continuous.*

Proof. We apply the closed graph theorem. Thus, let $x_n \to x$ in X and $Kx_n \to y$ in the F-space containing Y in the sense of Definition 5.1.1. We have to prove that $Kx = y$.
By passing to a subsequence, we may assume that $x_n \to x$ a.e., and that $|x_n| \leq |z|$ for some $z \in X$ (Corollary 3.2.1). Since K is absolutely defined, Lebesgue's dominated convergence theorem (with dominating function $s \mapsto |k(t,s)|\,|z(s)|$) implies that $Kx_n(t) \to Kx(t)$ for almost all $t \in T$. Now use the fact that Y is proper. □

We will see that the theorem may fail for ideal* spaces or if you drop the assumption of Y being proper (see Example 5.1.1).

We emphasize that the conditions of Theorem 5.1.2 do not imply that the operator (5.2) maps X into Y. This is not true even for $X = Y = L_2([0,1],\mathbb{R})$ [29, §4.3] (you may even choose K to be compact [25, Exercise 11.6], for details see [46, p. 38f]).

We also stress that in Theorem 5.1.2 for $x \in X$ the value $Kx(t)$ need not be defined for all t by (5.1) but only for almost all t. This is important for example, if you consider Volterra-operators

$$Kx(t) = \int_0^t k(t,s)x(s)ds$$

in spaces like $Y = C([0,1])$, where the kernel k is singular at 0: Here you would probably want to define $Kx(0) = \operatorname{ess\,lim}_{t\to 0+} Kx(t)$ (provided that the limit exists), hence the given operator might not be a 'pure' integral operator. Nevertheless, Theorem 5.1.2 may be applied.

Finally, we remark that Theorem 5.1.2 does not assume that the kernel k is measurable. For real kernels this is only of minor interest, but if $\mathfrak{L}(Z,W)$ is not separable, there are 'natural' integral operators, which are absolutely defined, but where k is not measurable (see Example 5.3.1).

Now we consider the case $T = S$ and the natural generalization of (5.1),

$$Ax(s) = c(s)x(s) + \int_S k(s,\sigma)x(\sigma)d\sigma, \tag{5.3}$$

where c and k take values in $\mathfrak{L}(Z,W)$. Again, we assume only that Ax is defined a.e.

If A maps a space X into another space Y, it need not happen that the two operators $Cx(s) = c(s)x(s)$ and $Kx(s) = \int_S k(s,\sigma)x(\sigma)d\sigma$ map X into Y also (although this is true of course under certain additional conditions, see e.g. [5]): To have an example you might choose $S = \mathbb{N}$ with the counting measure, $X = Y = l_\infty$, $c(s) = -s$, $k(s,s) = s$ and $k(s,\sigma) = 0$ for $s \neq \sigma$. For an example in the space of continuous functions, see e.g. [13].
The following theorem does not need this fact. For $X = Y = L_p([a,b])$ the proof can be found in [13] (for a similar proof, see also [26]):

Theorem 5.1.3. *Let X be an ideal space. Assume, the integral part of* (5.3) *is absolutely defined on X. If $A : X \to Y$, Y being some proper space, then A is continuous.*

Proof. By the closed graph theorem we have to show that $x_n \to x$ in X and $Kx_n \to y$ in the F-space containing Y in the sense of Definition 5.1.1 implies $y = Kx$. Corollary 3.2.1 implies that by passing to a subsequence we may assume that $x_n \to x$ a.e., and that $|x_n| \leq |z|$ for some $z \in X$. By Lebesgue's dominated convergence theorem we have $Kx_n \to Kx$ a.e., whence $Kx = y$, by the properness of Y. □

We remark that Theorem 5.1.3 need neither hold for a pure multiplication operator nor for a pure integral operator in Z-valued ideal* spaces, provided that Z is infinite-dimensional (and provided, of course, that we may use the axiom of choice):

Example 5.1.1. Let Z be an infinite dimensional Banach space with norm $|\cdot|$. As we have seen in Example 3.1.2, the axiom of choice implies that there exists another norm $|\cdot|^*$ on Z, which also turns Z in a Banach space, but such that there is a sequence $e_n \in Z$ with $|e_n| = 1$ and $|e_n|^* = n^{-1}$.
Now, let $S = \{0\}$, mes$S = 1$. Let X be the ideal* space of all functions $x : S \to Z$ with norm $\|x\|_X = |x(0)|^*$, and Y be the ideal space of all functions $x : S \to Z$ with norm $\|x\|_Y = |x(0)|$. Let $c(s) = k(t,s) = id$. Then both operators $Cx(s) = c(s)x(s)$ and $Kx(s) = \int_S k(t,s)x(s)ds$ are the identity operators from X to Y. However, although c and k take values in $\mathfrak{L}(Z)$, the identity is not a continuous operator from X into Y, because the sequence $x_n(0) = e_n$ satisfies $\|x_n\|_X = n^{-1} \to 0$, but $\|x_n\|_Y = 1$.

The previous example also shows that in Theorems 5.1.2 and 5.1.3 we may not drop the condition that Y is proper:
If we consider $|\cdot|^*$ as the natural norm on Z, then X becomes an ideal space, and the multiplication operator as well as the integral operator of course still are discontinuous mappings into Y (which now is an ideal* space). Observe that this implies by Theorem 5.1.3 that Y may not be proper.

For the multiplication operator Theorem 5.1.3 implies:

Theorem 5.1.4. *Let X be an ideal space, whose support has the finite subset property. Then for any measurable $c : S \to \mathfrak{L}(Z,W)$ (Z, W being Banach spaces) we have that $Cx(s) = c(s)x(s)$ maps X into itself, if and only if*

$$L = \operatorname*{ess\,sup}_{s \in \operatorname{supp} X} |c(s)| < \infty.$$

In this case we have $L = \|C\|$.

Proof. For sufficiency, observe that Cx is measurable by Theorem A.1.2, and that $\|C\| \leq L$. We now prove necessity. $\|C\| < \infty$ follows by Theorem 5.1.3. If $L > \|C\|$, there exists a number n, such that

$$M = \{s \in \operatorname{supp} X : |c(s)| \geq \|C\| + 2n^{-1}\}$$

is no null set. There exists some $y \in X$ where $E = \operatorname{supp} y$ is a subset of M with positive but finite measure. By Theorem A.2.1 there exists a measurable $z : E \to Z$, $|z| = 1$ with $|c(s)z(s)| \geq |c(s)| - n^{-1}$ on E. Put $x(s) = |y(s)|\, z(s)$ on E, $x(s) = 0$ outside E. Then $x \in X$, $x \neq 0$, and

$$|Cx(s)| \geq (|c(s)| - n^{-1})\, |y(s)| \geq (\|C\| + n^{-1})\, |x(s)| \qquad (s \in E),$$

which implies $\|Cx\| \geq (\|C\| + n^{-1})\, \|x\|$, a contradiction. □

If Z is finite-dimensional, Theorem 5.1.4 covers all multiplication operators:

Theorem 5.1.5. *Let X be a pre-ideal space. Let $c : S \to \mathfrak{L}(Z,W)$ (Z, W Banach spaces). Assume, $Cx(s) = c(s)x(s)$ is measurable for any $x \in X$. Then, if Z is finite-dimensional, c is measurable on each measurable subset of $\underline{\operatorname{supp}} X$.*

Proof. Otherwise, there exists a set $S \subseteq \underline{\operatorname{supp}} X$ of finite measure, such that c is on S not the limit of simple functions, i.e. not measurable on S. If we apply Corollary 2.2.7 to the restriction of X to functions vanishing outside S, we find that there exists a sequence of sets $E_1 \subseteq E_2 \subseteq \ldots$ with $\bigcup E_n = S$ and $\chi_{E_n} \in X$. If we can prove that c is measurable on any E_n, we have the contradiction that c is measurable on S.
Thus without loss of generality, we may assume that the underlying measure space has finite measure, and that X contains all constant functions, in particular the constant functions $x_n(s) \equiv e_n$ with $e_1, \ldots, e_N$ being a base of Z belong to X. Hence, by assumption, each function $y_n(s) = c(s)e_n$ is measurable, i.e. it is a.e. the limit of simple functions $x_k^n \to y_n$. Now define simple functions $c_k : S \to \mathfrak{L}(Z,W)$ by

$$c_k(s) \left[\sum_{n=1}^{N} z_n e_n \right] = \sum_{n=1}^{N} z_n x_k^n(s).$$

Then

$$| [c_k(s) - c(s)] \left[\sum_{n=1}^{N} z_n e_n \right] | \leq \sum_{n=1}^{N} |z_n| \, |x_k^n(s) - y_n(s)|.$$

But this implies $c_k \to c$ a.e., whence c is measurable. □

We remark however that for infinite-dimensional Z not all multiplication operators need to be generated by a measurable function, not even if $W = \mathbb{R}$:

Example 5.1.2. Let $S = [0,1]$, $Z = L_1(S)$, and $c : S \to Z^*$ be defined by $c(s)z = \int_0^s z(\sigma)d\sigma$. Then $Cx(s) = c(s)x(s)$ maps $X = L_1(S, Z)$ by Corollary A.1.1 into $Y = L_1(S, \mathbb{R})$, but c is not essentially separable valued, whence not measurable: This follows by $|c(s) - c(\sigma)| = 1$ for $s \neq \sigma$ and Lemma A.1.1.

We now consider continuity of linear partial integral operators. Let T, S, R be σ-finite measure spaces. Let X be a normed linear space of (classes of) measurable functions $S \times R \to Z$, Z being some Banach space. We say that the partial integral operator

$$Px(t, s) = \int_R k(t, s, \sigma)x(s, \sigma)d\sigma, \tag{5.4}$$

where the kernel k takes values in $\mathfrak{L}(Z, W)$, is *defined* on X, if for any $x \in X$ the function $\sigma \mapsto k(t, s, \sigma)x(s, \sigma)$ is integrable for almost all $(t, s) \in T \times S$. Again, we call two kernels k_1 and k_2 *equivalent* on X, if the corresponding partial integral operators P_1 and P_2 satisfy $P_1 x = P_2 x$ a.e. for any $x \in X$. Finally, if P is defined on X, it is called *absolutely defined* on X, if for some of its equivalent kernels the partial integral operator

$$|P| \, x(t, s) = \int_R |k(t, s, \sigma)| \, x(s, \sigma)d\sigma$$

is defined on X.

Again, for finite-dimensional Z this is no additional condition:

Theorem 5.1.6. *If X is a finite-dimensional-valued pre-ideal space, then any partial integral operator defined on X is absolutely defined on X.*

The proof follows that of Theorem 5.1.1.

Theorem 5.1.7. *Let X be an ideal space. Assume, the linear partial integral operator* (5.4) *is absolutely defined on X. If $P : X \to Y$, Y being some proper space, then P is continuous.*

The proof is almost the same as that of Theorem 5.1.2, see also [26].

Finally, we consider a generalization of [26]: Let $S_1, \ldots, S_n$ be σ-finite measure spaces, and let X be a normed linear space of functions $x : S_1 \times \ldots \times S_n \to Z$, Z being some Banach space. Let

$$Fx(s_1,\ldots,s_n) = c(s_1,\ldots,s_n)x(s_1,\ldots,s_n)+ \tag{5.5}$$
$$\sum_{k=1}^{n}\sum\int_{S_{i_1}\times\ldots\times S_{i_k}} f_{i_1\ldots i_k}(s_1,\ldots,s_n,\sigma_{i_1},\ldots,\sigma_{i_k})x(s_{i_1\ldots i_k})d(\sigma_{i_1},\ldots,\sigma_{i_k}).$$

Here the summation is taken over all indices $i_1 < \ldots < i_k$, $j_1 < \ldots < j_{n-k}$ with $\{i_1,\ldots,i_k\} \cup \{j_1,\ldots,j_{n-k}\} = \{1,\ldots,n\}$. Furthermore $s_{i_1\ldots i_k} = (t_1,\ldots,t_n)$ with $t_m = \sigma_m$ for $m \in \{i_1,\ldots,i_k\}$ and $t_m = s_m$ for $m \in \{j_1,\ldots,j_{n-k}\}$. The given functions $f_{i_1\ldots i_k}$ take values in $\mathfrak{L}(Z,W)$.
The operator (5.5) is in the same sense the natural generalization of a partial integral operator, as is (5.3) the natural generalization of an integral operator. Thus it is not surprising that we have the

Theorem 5.1.8. *Let X be an ideal space over $S_1 \times \ldots \times S_n$. Assume, the integral and partial integral operators of* (5.5) *are absolutely defined on X. If $F : X \to Y$, Y being some proper space, then F is continuous.*

The proof is almost the same than that of Theorem 5.1.3. see also [26].

5.2 Continuity and Uniform Continuity of Superposition Operators

We consider the following situation: Let S be some measure space, B and Z be Banach spaces, and let $D_f \subseteq S \times B$.

Definition 5.2.1. *A function $f : D_f \to Z$ is said to satisfy a* Carathéodory condition, *if $f(\cdot, u)$ is measurable for any u, and if $f(s, \cdot)$ is continuous for almost all $s \in S$.*

The measurability of $f(\cdot, u)$ and continuity of $f(s, \cdot)$ is of course only needed on the set of definition of the corresponding functions. Especially, if this set is empty, the condition is satisfied.

Given an $f : D_f \to Z$, we define the *superposition operator* $Fx(s) = f(s, x(s))$.
The next proposition is proved in [3] (see also [4]; for L_p-spaces see [27] or [29]). For σ-finite measure spaces observe that replacing the measure by the normalized measure (Corollary 2.2.6), does not touch the definition of X and Y, and that Y remains regular with respect to the new measure (Corollary 3.3.2).

Proposition 5.2.1. *Let $f : S \times \mathbb{R} \to \mathbb{R}$ satisfy a Carathéodory condition, where S is some σ-finite measure space. Assume, F maps some subset $D(F)$ of an ideal space X into a regular ideal space Y. Then F is continuous in the interior of $D(F)$.*

We want to prove this also for the vector-valued case. This is not as simple, as it seems to be, since the Krasnoselskiĭ-Ladyzhenskiĭ lemma [28] (see also [4, Theorem 6.2]) has no simple generalization. We prepare the proof by two easy lemmas, which are simple consequences of Lemma 3.3.1.

Lemma 5.2.1. *Let X be a pre-ideal* space, and x_n belong to the regular part of X. Let $D_1 \supseteq D_2 \supseteq \dots$ with $\bigcap D_n = \emptyset$. Then for any $\varepsilon > 0$ there exists a sequence $1 \leq n_1 < n_2 < \dots$ with*

$$||P_{D_{n_{k+1}}} x_{n_k}|| < \varepsilon \qquad (k \in \mathbb{N}).$$

Proof. Let $n_1 = 1$ and define the sequence by induction: If n_k is defined already, there exists an $n_{k+1} > n_k$ with $||P_{D_{n_{k+1}}} x_{n_k}|| < \varepsilon$ by Lemma 3.3.1. □

Lemma 5.2.2. *Let X be a pre-ideal* space. Let x_k belong to the regular part of X. Assume there exists a sequence of measurable sets $D_1 \supseteq D_2 \supseteq \dots$ with $\bigcap D_n = \emptyset$ and $\varepsilon > 0$ with*

$$\sup_k ||P_{D_n} x_k|| > \varepsilon \qquad (n \in \mathbb{N}).$$

Then there exists a sequence $1 \leq k_1 < k_2 < \ldots$ *with*

$$\|P_{D_n} x_{k_n}\| > \varepsilon.$$

Proof. We define the sequence by induction. Set $k_0 = 0$, and assume k_n is defined already. Since all x_k belong to the regular part of X, Lemma 3.3.1 implies that for $N \geq n+1$ large enough, we have

$$\|P_{D_N} x_k\| \leq \varepsilon \qquad (1 \leq k \leq k_n).$$

Hence, by assumption there is some $k_{n+1} > k_n$ with

$$\|P_{D_{n+1}} x_{k_{n+1}}\| \geq \|P_{D_N} x_{k_{n+1}}\| > \varepsilon.$$

□

The following theorem is proved by a tricky modification of the technique we already used in [46, Theorem 12.5]:

Theorem 5.2.1. *Let* $f : D_f \to Z$ *satisfy a Carathéodory condition. Let* X *be an ideal space, and* F *map some subset* $D(F) \subseteq X$ *into a regular ideal space* Y. *Assume that* Fx *has* σ*-finite support for any* $x \in D(F)$. *Then* F *is continuous in the interior of* $D(F)$.

Proof. Let x_0 be in the interior of $D(F)$. We have to prove that F is continuous at x_0. Putting $g(s,u) = f(s, u + x_0(s)) - Fx_0(s)$ and $Gx(s) = g(s, x(s))$, i.e. $Gx = F(x + x_0) - Fx_0$, it suffices to prove that G is continuous at 0. Assume, this is not the case. Since x_0 is in the interior of $D(F)$, we have that there exists some $r > 0$ such that $Gx \in Y$ for $\|x\| \leq r$. Since G is discontinuous at $G0 = 0$, there is some $\varepsilon > 0$ and a sequence $x_n \in X$ with $\|x_n\| \leq 2^{-n-1} r$ and $\|Gx_n\| > \varepsilon$.
By passing to a subsequence we may assume that $x_n \to 0$ a.e. (Corollary 3.1.2). Since then $Gx_n \to 0$ a.e., it suffices to prove by Theorem 3.3.3 that Gx_n has uniformly absolute continuous norms and vanishes uniformly at ∞ in norm.
Assume, this is not true. Since $\bigcup Gx_n$ has σ-finite support, Lemma 3.3.4 and Lemma 5.2.2 imply that there exists a sequence of sets $D_1 \supseteq D_2 \supseteq \ldots$ with $\bigcap D_n = \emptyset$, and a $\delta > 0$, such that (again passing to a subsequence of x_n if necessary)

$$\|P_{D_n} Gx_n\| > 2\delta \qquad (n \in \mathbb{N}).$$

Lemma 5.2.1 shows that, if we again pass to a subsequence of D_n and x_n, we may assume that we also have

$$\|P_{D_{n+1}} Gx_n\| < \delta \qquad (n \in \mathbb{N}).$$

Putting $E_n = D_n \setminus D_{n+1}$, we thus have

$$\|P_{E_n}Gx_n\| = \|P_{D_n}Gx_n - P_{D_{n+1}}Gx_n\| \geq \|P_{D_n}Gx_n\| - \|P_{D_{n+1}}Gx_n\| > \delta.$$

Theorem 3.2.1 and Corollary 3.2.2 imply that

$$z(s) = \sum_{n=1}^{\infty} P_{E_n}x_n(s)$$

belongs to X with

$$\|z\| \leq \sum_{n=1}^{\infty} \|P_{E_n}x_n\| \leq \sum_{n=1}^{\infty} \|x_n\| \leq r.$$

Hence, $Gz \in Y$ by assumption. Furthermore, since the E_n are pairwise disjoint, and since $G0 = 0$, we have

$$Gz(s) = \sum_{n=1}^{\infty} P_{E_n}Gx_n(s).$$

Since Gz belongs to the regular part of Y, Lemma 3.3.1 implies $\|P_{D_n}Gz\| \to 0$. But this is not possible, since

$$\|P_{D_n}Gz\| \geq \|P_{E_n}Gz\| = \|P_{E_n}Gx_n\| > \delta.$$

Thus we have found a contradiction. □

Observe that in the following corollary we do not assume that the underlying measure space is σ-finite:

Corollary 5.2.1. *If the superposition operator F generated by a Carathéodory function f maps some open set M of some ideal space X into $Y = L_p(S, Z)$, where $1 \leq p < \infty$, then F is continuous in M.*

Proof. Any function $y \in Y$ has σ-finite support: supp$y = \bigcup M_n$ with $M_n = \{s : |y(s)|^p \geq n^{-1}\}$, where $\infty > \|y\|_Y^p \geq n^{-1}\text{mes}M_n$. □

For regular spaces X, F is also continuous on the border of $D(F)$, provided the interior is not empty. We prepare this result by observing:

Corollary 5.2.2. *Let in the situation of Theorem 5.2.1 the interior of $D(F)$ contain some function y from the regular part of X. Let $M \subseteq D(F)$ vanish uniformly at ∞ in norm and have uniformly absolute continuous norms. Then $FM \subseteq Y$ has the same property, if it has σ-finite support.*

Proof. If the statement is false, Lemma 3.3.4 implies that there exists a sequence of sets $D_1 \supseteq D_2 \supseteq \ldots$ with $\bigcap D_n = \emptyset$, and a sequence $x_n \in M$ satisfying $\|P_{D_n} F x_n\| \not\to 0$. We now define $y_n = P_{D_n}(x_n - y) + y$. Then $\|y_n - y\| \le \|P_{D_n} x_n\| + \|P_{D_n} y\| \to 0$ by Lemma 3.3.3 and Lemma 3.3.1. Since F is continuous at y, and since $P_{D_n} F x_n = F y_n - F y + P_{D_n} F y$, Lemma 3.3.1 implies the contradiction $\|P_{D_n} F x_n\| \le \|F y_n - F y\| + \|P_{D_n} F y\| \to 0$. □

Corollary 5.2.3. *Let $f : D_f \to Z$ satisfy a Carathéodory condition. Assume, X, Y are regular ideal spaces, and F maps some subset $D(F) \subseteq X$ with nonempty interior into Y. Assume, furthermore, that Fx has σ-finite support for any $x \in D(F)$. Then F is continuous in $D(F)$.*

Proof. If the statement is false there is some $\varepsilon > 0$ and $x_n, x \in D(F)$ with $\|x_n - x\| \to 0$, but $\|F x_n - F x\| \ge \varepsilon$ for all n. By passing to a subsequence we may assume that $x_n \to x$ a.e. (Corollary 3.2.1). Then $F x_n \to F x$ a.e. If we can prove that the set of all $F x_n$ vanishes uniformly at ∞ in norm and has uniformly absolute continuous norms, we have a contradiction by Theorem 3.3.3. But the set M of all x_n has this property by Theorem 3.3.3. Thus we just have to apply Corollary 5.2.2, observing that FM has σ-finite support. □

In general, the assumptions of Theorem 5.2.1 do not imply that F is uniformly continuous on some ball of X, not even for $X = Y = L_p(S, \mathbb{R})$ [29, 17.6]. However, using Corollary 5.2.2, we can check a sufficient condition for F to be uniformly continuous, similar as [29, Theorem 17.4]. The generalization of that result is due to Chur-jen Chen, who has done this for the special spaces $[L_p \to L_q]$ in [10].

Since we want F to be uniformly continuous on some sets, we should at least assume that also $f(s, \cdot)$ is uniformly continuous on some sets. Thus, we suppose:

Definition 5.2.2. *A Carathéodory function f defined on some $D_f \subseteq S \times B$ is said to satisfy a* strict Carathéodory condition, *if for any $n \in \mathbb{N}$ and almost all $s \in S$ the function $f(s, \cdot)$ is uniformly continuous on*

$$D_f(s, n) = \{u : |u| \le n,\ (s, u) \in D_f\}.$$

If each $D_f(s, n)$ is compact, then any Carathéodory function f defined on D_f satisfies a strict Carathéodory condition. In particular, this is the case, if B is finite-dimensional, and almost all $\{u : (s, u) \in D_f\}$ are closed in B (e.g. if $D_f = S \times B$).

In general, the strict Carathéodory functions f are precisely the functions satisfying the generalized Hölder condition

$$|f(s,u)-f(s,v)| \leq \gamma_n(s,|u-v|) \qquad (|u|,|v| \leq n,\ (s,u),(s,v) \in D_f),$$

where $\gamma_n(s,t) \to 0$ as $t \to 0$ for any n and almost all s. Indeed, sufficiency is trivial, and for necessity define

$$\gamma_n(s,t) = \sup\{|f(s,u)-f(s,v)| : |u|,|v| \leq n,\ (s,u),(s,v) \in D_f,\ |u-v| \leq t\}.$$

Lemma 5.2.3. *Let M be a set of measurable functions on S. Assume, E has finite measure, such that M is bounded on E in measure, i.e.*

$$\lim_{n\to\infty} \sup_{x\in M} \{s \in E : |x(s)| \geq n\} = 0.$$

Let f be a strict Carathéodory function. Then F is on M uniformly continuous in measure on E, i.e. for any two sequences $x_n, y_n \in M$ with $x_n - y_n \to 0$ in measure on E, we have $Fx_n - Fy_n \to 0$ in measure on E (provided that Fx_n and Fy_n are defined).

Proof. Assume, the statement is false for some sequences $x_n, y_n \in M$. By passing to a subsequence, we may assume that there exist constants $\delta, \varepsilon > 0$, such that

$$E_n = \{s \in E : |Fx_n(s) - Fy_n(s)| \geq \delta\}$$

satisfies $\text{mes}E_n \geq \varepsilon$ for all n, and that $x_n - y_n \to 0$ a.e. on E. For N big enough, the sets

$$D_n = \{s \in E : |Fx_n(s)| > N \text{ or } |Fy_n(s)| > N\}$$

all satisfy $\text{mes}D_n < \varepsilon/2$. Since f satisfies a strict Carathéodory condition, the function $z_n = P_{E\setminus D_n}(Fx_n - Fy_n)$ converges to 0 on E a.e., whence also in measure. This implies that

$$C_n = \{s \in E \setminus D_n : |Fx_n(s) - Fy_n(s)| \geq \delta\}$$

satisfies $\text{mes}C_n < \varepsilon/2$ for n big enough. Now we have a contradiction by $E_n \subseteq C_n \cup D_n$. □

Observe that any bounded subset M of a pre-ideal space X satisfies the conditions of Lemma 5.2.3 (Corollary 3.1.3 for $y = \chi_E$).

We now have the following consequences:

Theorem 5.2.2. *Let the superposition operator F generated by some strict Carathéodory function f map some bounded subset M of a pre-ideal space X into some ideal space Y. Assume that for any countable $N \subseteq M$ the set $FN \subseteq Y$ has uniformly absolute continuous norms and vanishes uniformly at ∞ in norm. Then F is uniformly continuous on M.*

Proof. Let $x_n, y_n \in N$ with $\|x_n - y_n\| \to 0$. We will show that $\|Fx_n - Fy_n\| \to 0$. Let $\varepsilon > 0$ be given. The set of all Fx_n and all Fy_n vanishes uniformly at ∞ in norm. Thus, if we denote the underlying measure space by S, there exists a set $E \subseteq S$ of finite measure with

$$\sup_n \|P_{S\setminus E}(Fx_n - Fy_n)\| \leq \sup_n \|P_{S\setminus E}Fx_n\| + \sup_n \|P_{S\setminus E}Fy_n\| < \varepsilon.$$

By $\|Fx_n - Fy_n\| = \|P_E(Fx_n - Fy_n) + P_{S\setminus E}(Fx_n - Fy_n)\|$ it thus suffices to show that $z_n = P_E(Fx_n - Fy_n)$ satisfies $\|z_n\| \to 0$. By Lemma 5.2.3 and Theorem 3.1.1 we have $z_n \to 0$ in measure. Thus it suffices to apply Theorem 3.3.3, observing that the set of all z_n has uniformly absolute continuous norms, since so has the set of all Fx_n and Fy_n. □

The proof of the theorem was a bit technical, since we did not want to pose any assumptions on the underlying measure space.

Theorem 5.2.3. *Let the superposition operator F generated by some strict Carathéodory function f map some set $D(F)$ of an ideal space X into a regular ideal space Y. Assume, the interior of $D(F)$ contains a function in the regular part of X, and Fx has σ-finite support for any $x \in D(F)$. Let $M \subseteq D(F)$ be bounded such that any countable $N \subseteq M$ vanishes uniformly at ∞ in norm and has uniformly absolute continuous norms. Then F is uniformly continuous on M.*

Proof. The set FN has σ-finite support. Whence Corollary 5.2.2 implies that the conditions of Theorem 5.2.2 are satisfied. □

As an application of the previous theorems we have a slight generalization of [29, Theorem 17.4]:

Corollary 5.2.4. *Let S be a finite measure space, $1 \leq q < p \leq \infty$, $X = L_p(S,B)$, and Y be a regular ideal space over S, Assume, the superposition operator F generated by a strict Carathéodory function maps $L_q(S,B)$ into Y. Then $F : X \to Y$ is uniformly continuous on each (in X) bounded set $M \subseteq X$.*

Proof. Apply Theorem 5.2.3 for $\tilde{X} = L_q(S,B)$. Hölder's inequality implies that M has uniformly absolute continuous norms in $\tilde{X}$. □

Corollary 5.2.5. *Let S be a finite measure space, X be a pre-ideal space over S, $1 \leq p < q \leq \infty$, and $Y = L_p(S,Z)$. Assume, f satisfies a strict Carathéodory condition. Then for any bounded $M \subseteq X$ the mapping $F : M \to Y$ is uniformly continuous, provided that FM is defined and bounded in $L_q(S,Z)$.*

Proof. Apply Theorem 5.2.2. Hölder's inequality implies that FM has uniformly absolute continuous norms in Y. □

We remark that it is well-known that in the previous corollary for $S = [a, b]$, $X = L_r(S, \mathbb{R})$ $(1 \leq r \leq \infty)$ or, more general, X being a finite-dimensional valued Orlicz space over $[a, b]$, the mapping $F : X \to L_q(S, Z)$ is always bounded, if it is defined on X, i.e. the condition that FM is bounded is satisfied in this case (see [4] or [29], or [46] for the general case).

5.3 Continuity and Uniform Continuity of Hammerstein Operators

Now we give as an application of the previous theory a powerful condition for the continuity of Hammerstein operators, whose proof now is very easy.

We consider the following situation: Let T, S be σ-finite measure spaces, Y be a pre-ideal space over T, and Z, W be Banach spaces. Let $k : T\times S \to \mathfrak{L}(Z, W)$ be defined almost everywhere. We will not assume that k is measurable, but we will assume that the scalar function $|k|$ is measurable on $T \times S$. We denote the largest normed linear space, which is mapped by the integral operator with kernel $|k|$, by $Y[k]$:

Definition 5.3.1. *Let $Y[k]$ be the space of all measurable $x : S \to Z$, where*

1. *$x(s) = 0$ whenever $k(\cdot, s) = 0$ a.e., and*
2. *the function*

$$|K|\,|x|\,(t) = \int_S |k(t,s)|\,|x(s)|\,ds \tag{5.6}$$

is finite a.e. and belongs to Y.

Define $\|x\|_{Y[k]} = \|\ |K|\,|x|\ \|_Y$.

$Y[k]$ is in fact a normed linear space:

Lemma 5.3.1. *$Y[k]$ is a pre-ideal space. If Y is regular, then $Y[k]$ is regular. If Y is almost α-perfect, then $Y[k]$ is almost α-perfect. If Y is semi-perfect, then $Y[k]$ is a semi-perfect ideal space.*

Proof. $Y[k] = [L_1 \to Y]|_S^{|k|}$. Use Corollary 4.1.1, and Theorems 4.1.6, 4.1.7, 4.1.8, and 4.2.2. □

Since we did not assume that k is measurable, it is in general not true that the linear integral operator

$$Kx(t) = \int_S k(t,s)x(s)ds \tag{5.7}$$

is defined on $Y[k]$. However, it is easy to see that its domain of definition (i.e. the set of all x, for which Kx is defined a.e. and measurable) is a linear space $D(K)$. We equip the linear space $D(K) \cap Y[k]$ with the norm of $Y[k]$.

Lemma 5.3.2. (5.7) *is a continuous operator from $D(K)\cap Y[k]$ into Y with* $\|K\| \leq 1$.

Proof. For $x \in D(K) \cap Y[k]$ the equation $|Kx(s)| \leq |K|\,|x|\,(s)$ implies $Kx \in Y$ and $\|Kx\|_Y \leq \|\ |K|\,|x|\ \|_Y = \|x\|_{Y(K)}$. □

Now we have reached our goal:

Theorem 5.3.1. *Let T, S be σ-finite measure spaces, B, Z, W be Banach spaces, and $D_f \subseteq S \times B$. Let $f : D_f \to Z$ satisfy a Carathéodory condition, and $k : T \times S \to \mathfrak{L}(Z, W)$ be defined a.e., such that $|k|$ is measurable. Assume,*

$$|H|\, x(t) = \int_S |k(t,s)|\, |f(s, x(s))|\, ds$$

maps some subset $D(|H|)$ of an ideal space X into the completely regular ideal space Y. Then the Hammerstein operator

$$Hx(t) = \int_S k(t,s) f(s, x(s)) ds$$

is continuous at each point of its set of definition, which belongs to the interior of $D(|H|)$.

Proof. Without loss of generality we may assume that $f(s, \cdot) = 0$, if $k(\cdot, s)$ vanishes almost everywhere. Then the superposition operator $Fx(s) = f(s, x(s))$ maps $D(|H|)$ into $Y[k]$, hence is continuous in the interior of $D(|H|)$ by Theorem 5.2.1 and Lemma 5.3.1. Since $H = K \circ F$, we are done by Lemma 5.3.2. □

Analogously, by using Corollary 5.2.3 instead of Theorem 5.2.1 we get

Theorem 5.3.2. *Let in Theorem 5.3.1 also X be regular, and $D(|H|)$ have nonempty interior. Then H is continuous on $D(|H|)$, wherever defined.*

Similar, Theorem 5.2.3 yields:

Theorem 5.3.3. *Let in Theorem 5.3.1 the interior of $D(|H|)$ contain a function in the regular part of X, and f satisfy a strict Carathéodory condition. Then H is uniformly continuous on each (in X) bounded subset $M \subseteq D(|H|)$, which vanishes uniformly at ∞ in norm and has uniformly absolute continuous norm (wherever defined on M).*

Corollary 5.3.1. *If f satisfies a strict Carathéodory condition, S has finite measure, $1 \leq q < p \leq \infty$, and $|H|$ is defined on $L_q(S, B)$ with values in a regular space Y, then for $X = L_p(S, B)$, $D(H) \subseteq X$, the mapping $H : D(H) \to Y$ is uniformly continuous on each (in X) bounded set $M \subseteq D(H)$.*

Proof. Apply Theorem 5.3.3 for $\tilde{X} = L_q(S, B)$. By Hölder's inequality, M has uniformly absolute continuous norm in $\tilde{X}$. □

We took much care in not assuming that the kernel k is measurable. This seems to be quite artificial, if you think of k as a real function. But if $\mathfrak{L}(Z, W)$ is not separable, measurability of the kernel is very restrictive even for 'natural' integral operators K:

Example 5.3.1. Let $T = S = [0,1]$, $Z = L_1(S, \mathbb{R})$, $W = \mathbb{R}$. Let $k(t,s) = c(s)$, where $c : S \to Z^*$ is defined by $c(s)z = \int_0^s z(\sigma)d\sigma$. As we have seen in Example 5.1.2, c is not essentially separable valued, but the multiplication operator $Cx(s) = c(s)x(s)$ maps $L_1(S,Z)$ into $L_1(S,W)$. This means that k is not measurable, but that (5.7) is defined on e.g. $X = L_p(S,Z)$ $(1 \leq p \leq \infty)$ and maps X into e.g. $Y = L_q(T,W)$ $(1 \leq q \leq \infty)$.

We remark that by $|k(t,s)| = 1$ it is easy to see in this example that K is also absolutely defined on X, and even more that the operator (5.6) maps X into Y.

5.4 Some Applications to Barbashin Equations

In this section we present some applications of the previous theory to integro-differential equations of Barbashin type and to partial integral operators. Emphasis is put on the reformulation of the problems in appropriate ideal spaces.

At first, we consider the most simple case of a linear Barbashin equation, namely

$$\frac{\partial x(t,s)}{\partial t} = c(t,s)x(t,s) + \int_S k(t,s,\sigma)x(t,\sigma)d\sigma + f(t,s). \tag{5.8}$$

Here, $(t,s) \in I \times S$ with I being some interval. S is a σ-finite measure space, usually a bounded interval $[a,b]$. c, k and f are given functions (for simplicity, we assume they are product-measurable), and x is the unknown function. We explicitly allow the case that x takes values in some Banach space Y (and thus also f, while c and k take values in $\mathfrak{L}(Y)$), which can be interpreted in the way that we even deal with a finite (in case $Y = \mathbb{R}^n$) or even infinite (e.g. in case $Y = l_p$) system of equations of the form (5.8).
To bring this in a functional analytic setting, it seems natural to consider the abstract function $x : I \to X$ with some appropriate ideal space X, and to identify $x(t)(s) = x(t,s)$. To make this more precise, we recall (see Sect. 4.4) that $W_m(I,X)$ is the space of all measurable functions $x : I \times S \to Y$, such that $x(\cdot,s)$ satisfies FTC for all s, and such that $(t,s) \mapsto \partial x(t,s)/\partial t$ is measurable, and $t \mapsto \partial x(t,\cdot)/\partial t$ is locally integrable as an abstract function $I \to X$. If this abstract function is even continuous, we say $x \in C_m^1(I,X)$. Recall that for finite-dimensional Y the function $x(\cdot,s)$ satisfies FTC, if and only if it is absolutely continuous.

Definition 5.4.1. *A function x is called a* solution *of* (5.8) *in X, if $x \in C_m^1(I,X)$, and if for any $t \in I$ we have that* (5.8) *holds for almost all s. A function x is called* generalized solution *of* (5.8) *in X, if $x \in W_m(I,X)$, and if* (5.8) *holds for almost all $(t,s) \in I \times S$.*

Now, we consider the operator

$$A(t)u(s) = c(t,s)u(s) + \int_S k(t,s,\sigma)u(\sigma)d\sigma$$

and the abstract function $f : t \mapsto f(t,\cdot)$. It seems to be natural to rewrite (5.8) as a differential equation in the ideal space X,

$$\frac{dx(t)}{dt} = A(t)x(t) + f(t). \tag{5.9}$$

Theorem 4.4.4 and Corollary 4.4.3 imply that the two equations are equivalent in the following sense:

Theorem 5.4.1. *Let X be an ideal space. Assume, for almost all t we have*

$$A(t)u + f(t) \in X \qquad (u \in X). \tag{5.10}$$

Then the generalized solutions x of (5.8) in X are precisely the functions $x : I \to X$ satisfying FTC, which solve (5.9) for almost all t.
If (5.10) even holds for all t, then the solutions x of (5.8) in X are precisely the continuously differentiable solutions $x : I \to X$ of (5.9).

The particular choice of X will in practice mainly depend on the condition (5.10).

By Theorem 5.4.1 we may now apply well-known results for the differential equation (5.9) in the Banach space X (see e.g. [11]). For example, we immediately get:

Theorem 5.4.2. *Let A be locally integrable as a function $T \to \mathfrak{L}(X)$, and f be locally integrable as a function $T \to X$. Then for any $t_0 \in I$, $x_0 \in X$ the equation (5.8) has precisely one generalized (local and global) solution x in X which satisfies the initial value condition $x(t_0, \cdot) = x_0$ a.e.*
If additionally, f is continuous, and A is strongly continuous (i.e. $t \mapsto A(t)u$ is continuous for any $u \in X$), then x is a solution in the sense of Definition 5.4.1.

We emphasize that our reasoning did not depend on the particular form of A. For Theorem 5.4.1 we did not even need that $A(t)$ is linear. Thus this theorem is applicable e.g. for the nonlinear Barbashin equation

$$\frac{\partial x(t,s)}{\partial t} = c(t,s)x(t,s) + \int_S k(t,s,\sigma,x(t,\sigma))d\sigma + f(t,s),$$

which was used implicitly in [16]. Similarly, Theorem 5.4.2 also may be formulated for generalized linear Barbashin equations of the form

$$\begin{aligned}\frac{\partial x(t,\tau,s)}{\partial \tau} &= c(t,\tau,s)x(t,\tau,s) \\ &\quad + \int_S l(t,\tau,s,\sigma)x(t,\tau,\sigma)d\sigma + \int_T m(t,\tau,s,r)x(r,\tau,s)dr \\ &\quad + \int_T\int_S n(t,\tau,s,\sigma,r)x(r,\tau,\sigma)d\sigma\, dr + f(t,\tau,s),\end{aligned}$$

which was done in [9] (silently using our proofs).

As an example for a simple nonlinear application we will consider the initial value problem for a Barbashin equation with degenerated nonlinearity:

$$
\begin{aligned}
\frac{\partial x(t,s)}{\partial t} &= c(t,s)x(t,s) + \int_S k(t,s,\sigma)x(t,\sigma)d\sigma \\
&\quad + \sum_{i=1}^{n} \int_S k_i(s) f_i(\sigma, x(t,\sigma))d\sigma + f(t,s), \qquad (5.11)
\end{aligned}
$$

$$x(t_0, \cdot) = x_0.$$

We consider the operators

$$
\begin{aligned}
A(t)u(s) &= c(t,s)u(s) + \int_S k(t,s,\sigma)u(\sigma)d\sigma, \\
F_i u &= \int_S f_i(\sigma, u(\sigma))d\sigma, \text{ and} \\
K_i u(s) &= \int_S k_i(s) f_i(\sigma, u(\sigma))d\sigma = k_i(s) F_i u.
\end{aligned}
$$

Besides of the condition of Theorem 5.4.2, we will assume that the f_i are real-valued, and that the K_i are uniformly continuous on bounded subsets of X (which might e.g. be checked by Theorem 5.3.3). We will prove that then (5.11) has a local (generalized) solution in X (this is not trivial, since we do not put any Lipschitz-condition on F_i).
The assumptions imply that $k_i \in X$, and thus that $F_i u$ is defined on X. We consider the integrated equation in X

$$x(t) = Bx(t) + Kx(t) + G(t), \qquad (5.12)$$

where

$$
\begin{aligned}
Bx(t) &= \int_{t_0}^{t} A(\tau)x(\tau)d\tau, \\
Kx(t) &= \int_{t_0}^{t} \sum K_i x(\tau)d\tau = \sum \left(\int_{t_0}^{t} F_i x(\tau)d\tau \right) k_i, \text{ and} \\
G(t) &= \int_{t_0}^{t} f(\tau)d\tau + x_0.
\end{aligned}
$$

We have to show that there exists a continuous solution of (5.12). Without loss of generality, let I be a compact interval. Then we may consider (5.12) as a fixed-point equation in the space Y of continuous mappings $I \to X$ with the maximum-norm. It is well-known that $id - B$ is continuously invertible in Y (see e.g. [11]). Thus (5.12) may be written in Y as the fixed point equation

$$x = (id - B)^{-1}(Kx + G) =: Mx.$$

We will prove by Schauder's theorem that M has a fixed point in a sufficiently small ball U around $(id - B)^{-1}G = G_0$, provided that the intervall I is just small enough: By the uniform continuity of K_i the size $||Kx(t)||$ becomes

uniformly small for all t near t_0 and all x near e.g. G_0. Thus $MU \subseteq U$ can be ensured by a proper choice of U and small I.
It suffices to prove now that K is completely continuous. But this follows by the theorem of Arzelà-Ascoli: For any fixed t the set $\{Kx(t) : x \in U\}$ is bounded and contained in the linear hull of $b_1, \ldots, b_n$, whence this set is precompact in X. And a standard calculation shows that the uniform continuity of K_i implies that K is continuous on U, and that KU is equicontinuous.

Of course, the same idea may be used for the more general equation

$$\begin{aligned}\frac{\partial x(t,s)}{\partial t} &= c(t,s)x(t,s) + \int_S k(t,s,\sigma)x(t,\sigma)d\sigma \\ &+ \int_S p(t,s,\sigma,x(t,\sigma))d\sigma + f(t,s),\end{aligned}$$

where we now have

$$\begin{aligned}Kx(t) &= \int_{t_0}^{t} P(\tau)x(\tau)d\tau \text{ with} \\ P(t)u &= \int_S p(t,s,\sigma,u(\sigma))d\sigma,\end{aligned}$$

if we can check somehow e.g. that K is defined and completely continuous on U, and that $MU \subseteq U$. Since K (if it is defined) after our usual identification of ideal-valued functions may be interpreted as a Uryson operator of Volterra-type

$$Kx(t,s) = \int_{t_0}^{t} \int_S p(\tau,s,\sigma,x(\tau,\sigma))d\sigma\, d\tau,$$

this can be done in many cases.

We now consider the simplest form of partial integral operators [26],

$$Lx(t,s) = \int_S l(t,s,\sigma)x(t,\sigma)d\sigma, \tag{5.13}$$

where l is product-measurable for simplicity. We are looking for 'natural' ideal spaces X, Y such that L maps X into Y (continuously, cf. Sect. 5.1). For this purpose, we consider the family of integral operators

$$L(t)u(s) = \int_S l(t,s,\sigma)u(\sigma)d\sigma.$$

We assume that for almost all t we have found ideal spaces $U_1(t), U_2(t)$ such that $L(t)$ maps $U_1(t)$ (continuously) into $U_2(t)$, with

$$\|L(t)\|_{\mathfrak{L}(U_1(t),U_2(t))} = c(t).$$

Let us assume that the families $U_i(\cdot)$ are T-measurable and that c is measurable. Now, let V_1, V_2 be two ideal spacs such that the multiplication operator $Cv(t) = c(t)v(t)$ maps V_1 (continuously) into V_2,

$$\|C\|_{\mathfrak{L}(V_1,V_2)} = N.$$

Then (5.13) maps $X = [U_1(\cdot) \to V_1]$ into $Y = [U_2(\cdot) \to V_2]$, and $\|L\| \le N$, since

$$\begin{aligned} \|Lx\|_Y &= \|t \mapsto \|L(t)x(t,\cdot)\|_{U_2(t)}\|_{V_2} \le \|t \mapsto c(t)\,\|x(t,\cdot)\|_{U_1(t)}\|_{V_2} \\ &\le N\|t \mapsto \|x(t,\cdot)\|_{U_1(t)}\|_{V_1} = N\,\|x\|_X\,. \end{aligned}$$

Usually, we even have $\|L\| = N$, and the assumptions are 'sharp': At first, for any $\varepsilon > 0$ there exists some $v \in V_1$, $\|v\|_{V_1} \le 1$ such that

$$\|Cv\|_{V_2} \ge N - \varepsilon.$$

Similarly, for almost all t there exists some $u_t \in U_1(t)$, $\|u_t\|_{U_1(t)} \le 1$ satisfying

$$\|L(t)u_t\|_{U_2(t)} \ge c(t)(1 - \varepsilon).$$

We assume now that the function $x(t,s) = |v(t)|\,|u_t(s)|$ is measurable on $T \times S$ (for $T = \mathbb{N}$ this is always satisfied; for uncountable T you even have to use the axiom of choice in general to just assure that x is a function; in many cases – e.g. if the $U_i(\cdot)$ are countable measurable and l is nonnegative and independent of s – you may apply Theorem A.3.4; observe that the problem of 'constructing' such a measurable function x is similar to the problem discussed in the proofs of the Theorems 4.3.7 and 4.3.8).
However, if this assumption is satisfied, x belongs to the real form of X,

$$\|x\|_X = \|t \mapsto |v(t)|\ \|u_t\|_{U_1(t)}\|_{V_1} \le \|v\|_{V_1} \le 1,$$

and (without loss of generality let $\varepsilon < 1$)

$$\begin{aligned} \|Lx\|_Y &= \|t \mapsto |v(t)|\ \|L(t)u_t\|_{U_2(t)}\|_{V_2} \ge \|t \mapsto |v(t)|\,c(t)(1-\varepsilon)\|_{V_2} \\ &= (1-\varepsilon)\,\|Cv\|_{V_2} \ge (1-\varepsilon)(N-\varepsilon). \end{aligned}$$

For $\varepsilon \to 0$ this implies $\|L\| \ge N$.

Thus we have seen that the spaces with mixed family norm are the 'natural' spaces to consider the operator (5.13). In [26] the operator (5.13) is considered in spaces with mixed norm (also the norm estimate $\|L\| \le N$ is proved there for this case).

A. Appendix: Some Measurability Results

A.1 Sup-Measurable Operator Functions

Let S be some measure space, and Y, Z be Banach spaces. Let $A(s) : Y \to Z$ be an (not necessarily linear) operator function for almost all $s \in S$.
One often has the problem to integrate the function $s \mapsto A(s)x(s)$, where $x : S \to Y$. Boundedness is usually trivial, thus it suffices to check measurability:

Definition A.1.1. *The operator function A is called* sup-measurable, *if $s \mapsto A(s)x(s)$ is measurable for any measurable $x : S \to Y$.*

Theorem A.1.1. *Let for almost all s the mapping $A(s) : Y \to Z$ be continuous. Then A is sup-measurable, if and only if $s \mapsto A(s)y$ is measurable for any $y \in Y$.*

Proof. We prove sufficiency. First, assume $x(s) = \sum x_k \chi_{E_k}(s)$ is simple, where E_k are pairwise disjoint with $\bigcup E_k = S$. Since each $y_k(s) = A(s)x_k$ is measurable, also $\sum \chi_{E_k}(s) y_k(s) = A(s)x(s)$ is measurable.
In the general case, let $E \subseteq S$ have finite measure. Choose a sequence of simple functions $x_n : S \to Y$, which converges to x a.e. on E. Then $A(s)x_n(s) \to A(s)x(s)$ for almost all $s \in E$. Thus, $s \mapsto A(s)x(s)$ is measurable on E, whence measurable. □

In different terminology the main part of Theorem A.1.1 may be reformulated as: The superposition operator generated by the Carathéodory function $f(s, y) = A(s)y$ maps measurable functions into measurable functions. This is of course well-known for scalar y (see e.g. [4]).

Recall that an operator function A is said to be *strongly continuous*, if $s \mapsto A(s)y$ is continuous for any fixed y.

Corollary A.1.1. *Let I be some compact interval, $A : I \to \mathfrak{L}(Y, Z)$ be a strongly continuous operator function. Then for any integrable $x : I \to Y$ also $s \mapsto A(s)x(s)$ is integrable.*

Proof. A is sup-measurable, since $s \mapsto A(s)y$ is continuous for all $y \in Y$. By the uniform boundedness principle, $\|A(s)\| \leq C$ for all $s \in I$. Thus

$\|A(s)x(s)\| \le C\,\|x(s)\|$. □

We remark that in Corollary A.1.1 it may happen that A is not measurable, even if $Y = Z$ is separable:

Example A.1.1. Let $I = [0,1]$, and $Y = Z = L_1(I)$. Define $A : I \to \mathfrak{L}(Y)$ by $A(s)x(t) = \int_0^s x(\sigma)d\sigma$. A is strongly continuous but not measurable, since it is not essentially separable valued by $|A(s) - A(\sigma)| = 1$ for $s \neq \sigma$ and the following Lemma A.1.1.

We remark that the proof of the following lemma is straightforward, if the axiom of choice may be used. But since we just want to use the principle of dependent choices, we have to argue in a different way:

Lemma A.1.1. *Let X be a metric space, $M \subseteq X$. Assume, there exist $\delta > 0$ and an uncountable $U \subseteq M$ with*

$$d(x,y) \ge \delta \qquad (x, y \in U,\ x \neq y).$$

Then M is not separable in X.

Proof. If M is separable in X, there exists an at most countable set $E \subseteq X$, whose closure contains M, hence also U. Let D consist of all $n \in E$ with $\operatorname{dist}(n, U) < \delta/2$. Then also D is at most countable, and its closure contains U.
For any $n \in D$ choose some $x_n \in U$ with $d(n, x_n) < \delta/2$. Then the set C of all x_n is at most countable. Hence there exists some $y \in U \setminus C$. For any $n \in D$ we have $\delta \le d(x_n, y) \le d(x_n, n) + d(n, y) \le \delta/2 + d(n, y)$, hence y does not belong to the closure of D, a contradiction. □

Thus even in the special case of Example A.1.1 sup-measurability does not imply measurability in operator norm (although this is true for $A : S \to \mathfrak{L}(Y, Z)$ with finite-dimensional Y: Apply Theorem 5.1.5 for $X = L_\infty(S, Y)$ and $c = A$). The converse usually is true:

Theorem A.1.2. *Let Y, Z be Banach spaces, and W be a Banach space of continuous mappings $Y \to Z$. Assume, $A_n \to A$ in W implies that for all $y \in Y$ we have $A_n y \to Ay$. Then any measurable $A : S \to W$ is sup-measurable.*

Proof. Let $A : S \to W$ be measurable. Let $y \in Y$. Let $E \subseteq S$ have finite measure. Choose a sequence of simple functions $A_n : S \to W$, converging to A a.e. on E. Then $A_n(s)y \to A(s)y$ for almost all $s \in E$. Since each $s \mapsto A_n(s)y$ is also simple, $s \mapsto A(s)y$ is measurable on E, whence measurable. Now use Theorem A.1.1. □

A.2 Majorising Principles for Measurable Operator Functions

We are now going to prove for a measurable operator function A the existence of a measurable function x, such that $s \mapsto A(s)x(s)$ has some majorising properties. The problem consists in finding a *measurable* such x.

The main tool is the following lemma, which is quite an immediate consequence of Egorov's theorem:

Lemma A.2.1. *Let Y be a Banach space. Then for any measurable $x : S \to Y$ with σ-finite support and any measurable function y with* $\operatorname{supp} x \subseteq \operatorname{supp} y$ *there exists a sequence of measurable sets $S_n \subseteq S$ with* $\operatorname{mes}(S \setminus \bigcup S_n) = 0$ *and a sequence x_n of simple functions with*

$$|x(s) - x_n(s)| \leq |y(s)| \qquad (s \in S_n). \tag{A.1}$$

Moreover, we may additionally satisfy $S_1 \subseteq S_2 \subseteq \ldots$ and $x_k|_{S_k} = x_n|_{S_k}$ for $k \leq n$.

Proof. 1. We first drop the additional assumptions. Since we may put $x_n(s) = 0$ for $s \notin \operatorname{supp} x$, we may assume without loss of generality that $S = \operatorname{supp} x$. Let $S = \bigcup E_n$, where $E_1 \subseteq E_2 \subseteq \ldots$ have finite measure. Choose any sequence z_k of simple functions, which converges a.e. to x. Now fix n. Since $E_n \subseteq \operatorname{supp} y$, there exists some N such that the measure of $M = \{s \in E_n : |y(s)| \leq N^{-1}\}$ is less that n^{-1}. By Egorov's theorem there exists a measurable subset $S_n \subseteq E_n \setminus M$ with $\operatorname{mes}(E_n \setminus S_n) < 2n^{-1}$, such that $z_k \to x$ uniformly on S_n. Hence some $x_n = z_{k_n}$ satisfies $|x(s) - x_n(s)| \leq N^{-1}$ for $s \in S_n$. Thus $S_n \cap M = \emptyset$ implies (A.1). Moreover, each $Q_n = E_n \setminus \bigcup_k S_k$ is a null set, since for each $k \geq n$ we have $Q_n \subseteq E_k \setminus S_k$. Hence $\bigcup Q_n = S \setminus \bigcup S_k$ is a null set.
2. By 1. there exists a sequence of measurable sets T_n with $\operatorname{mes}(S \setminus \bigcup T_n) = 0$, and a sequence of simple functions y_n with $|x(s) - y_n(s)| \leq |y(s)|$ for $s \in T_n$. Now put

$$E_n = T_n \setminus \bigcup_{k<n} T_k, \; S_n = \bigcup_{k \leq n} E_k = \bigcup_{k \leq n} T_k.$$

Since the sets E_n are pairwise disjoint, the simple functions

$$x_n(s) = \sum_{k=1}^{n} \chi_{E_k}(s) y_k(s)$$

satisfy (A.1). The additional assumptions are also satisfied. □

With this lemma we may prove the first result:

Theorem A.2.1. *Let Y be a normed linear space, Z be a Banach space, and S be a measure space. Assume, $A : S \to \mathfrak{L}(Y, Z)$ is measurable with σ-finite support. Then to any measurable y with* $\mathrm{supp} A \subseteq \mathrm{supp} y$ *and to any nonempty bounded set $W \subseteq Y$ there exists a measurable and essentially countable-valued function $x : S \to W$ with*

$$|A(s)x(s)| \geq |A(s)|_W - |y(s)| \qquad (s \in S),$$

where $|A|_W = \sup_{w \in W} |Aw|$.

Proof. 1. First assume additionally that $S = \mathrm{supp} A$ has finite measure and that $|y(s)| \geq 3\varepsilon > 0$ on S.
By Lemma A.2.1 there exists a sequence of measurable sets $S_n \subseteq S$ with $\mathrm{mes}(S \setminus \bigcup S_n) = 0$ and a sequence of simple functions $A_n : S \to \mathfrak{L}(Y, Z)$ with

$$|A(s) - A_n(s)|_W \leq \varepsilon \qquad (s \in S_n).$$

Since we may additionally assume that $S_1 \subseteq S_2 \subseteq \ldots$ and $A_k|_{S_k} = A_n|_{S_k}$ for $k \leq n$, we may assume without loss of generality that

$$A_n(s) = \sum_{k=1}^{m_n} \chi_{D_k}(s) A_k,$$

where the pairwise disjoint measurable sets D_k and $A_k \in \mathfrak{L}(Y, Z)$ are independent of n with $\bigcup D_k = \bigcup S_n$ and $m_1 < m_2 < \ldots$. For any k there exists some $w_k \in W$, with $|A_k|_W \leq |A_k w_k| + \varepsilon$. Now define

$$x(s) = \sum_{k=1}^{\infty} \chi_{D_k}(s) w_k.$$

Then x has the properties $x(s) \in W$ a.e. and $|A_n(s)|_W \leq |A_n(s)x(s)| + \varepsilon$ for $s \in S_n$. Particularly, for $s \in S_n$ we have

$$|A(s)x(s)| \geq |A_n(s)x(s)| - |\, [A_n(s) - A(s)]\, x(s)| \geq |A_n(s)|_W - 2\varepsilon \geq |A(s)|_W - 3\varepsilon$$

Since $\mathrm{mes}(S \setminus \bigcup S_n) = 0$, and $|y(s)| \geq 3\varepsilon$, this implies the statement (by modifying x on a proper null set).
2. Let $\mathrm{supp} A = \bigcup S_n$, where S_n have finite measure, and put

$$T_{nk} = \{s \in S_n : |y(s)| \geq k^{-1}\}.$$

By 1. there exist measurable and essentially countable-valued functions $x_{nk} : T_{nk} \to W$ with $|A(s)|_W \leq |A(s)x(s)| + |y(s)|$ for $s \in T_{nk}$. Let $M : \mathbb{N} \to \mathbb{N} \times \mathbb{N}$ be onto, and define

$$E_n = T_{M(n)} \setminus \bigcup_{k<n} T_{M(k)}.$$

For $s \in E_n$ put $x(s) = x_{M(n)}(s)$, and for $s \in S \setminus \bigcup E_n$ put $x(s) = w$, where $w \in W$ is fixed. Since the E_n are pairwise disjoint, x is well defined, measurable and essentially countable-valued. We have $x(s) \in W$, and for $s \in E = \bigcup E_n = \bigcup_{nk} T_{nk}$ we have $|A(s)|_W \leq |A(s)x(s)| + |y(s)|$. Thus, if we modify x on the null set $(\operatorname{supp} A) \setminus E$, we have the statement. □

We emphasize that the following consequence holds also for complex Banach spaces:

Corollary A.2.1. *Let Y be a Banach space, Y^* be its dual space, and S be a measure space. Let $y : S \to Y^*$ be measurable. Then for any measurable z, w with σ-finite $M = \operatorname{supp} y \cap \operatorname{supp} w$ and $M \subseteq \operatorname{supp} z$, there exists a measurable $x : S \to Y$ with $|x(s)| = |w(s)|$, $y(s)x(s) \geq 0$, and*

$$y(s)x(s) \geq |y(s)|\,|w(s)| - |z(s)| \qquad (s \in S).$$

Proof. Since outside M you may define $x(s) = w(s)\operatorname{sgn}[\overline{y(s)w(s)}]$ (for complex $z \neq 0$ put $\operatorname{sgn} z = z/|z|$), we may assume without loss of generality that $M = S$. Theorem A.2.1 implies for $A = y$ that there exists a measurable $u : S \to Y$ with $|u| = 1$ and

$$|y(s)u(s)| \geq |y(s)| - \min\{|w(s)|^{-1}\,|z(s)|\,, |y(s)|\}.$$

Now just put $x(s) = u(s)\,|w(s)|\operatorname{sgn}[\overline{y(s)u(s)}]$. □

For $w \equiv 1$ and $z \equiv \varepsilon$ an analogous result may be found in [51, Lemma 2] (for finite measure spaces).

The 'dual' version of Corollary A.2.1 uses the fact that the canonical embedding of Y into Y^{**} is norm-preserving (see Definition 2.1.8).

Corollary A.2.2. *Let Y be a Banach space with the bidual property, Y^* be its dual space, and S be a measure space. Let $x : S \to Y$ be measurable. Then to any measurable z, w with σ-finite $M = \operatorname{supp} x \cap \operatorname{supp} w$ and $M \subseteq \operatorname{supp} z$, there exists a measurable $y : S \to Y^*$ with $|y(s)| = |w(s)|$, $y(s)x(s) \geq 0$, and*

$$y(s)x(s) \geq |w(s)|\,|x(s)| - |z(s)| \qquad (s \in S).$$

Proof. Let $i : Y \to Y^{**}$ be the canonical embedding, and let $u = i \circ x$, i.e. $u(s)l = l(x(s))$. u is measurable, since i is continuous. Corollary A.2.1 applied for Y^* instead of Y and for u instead of y yields that there exists a measurable $y : S \to Y^*$ with $|y(s)| = |u(s)|$, $u(s)y(s) \geq 0$, and

$$u(s)y(s) \geq |u(s)|\,|w(s)| - |z(s)| \qquad (s \in S).$$

This implies the statement, since $u(s)y(s) = y(s)x(s)$, and since $|u(s)| = |x(s)|$, because Y has the bidual property. □

Remark A.2.1. If we assume the axiom of choice and are just looking for a measurable (i.e. not necessarily essentially countable-valued) function x, the condition of $M = \operatorname{supp} A$ being σ-finite in Theorem A.2.1 (and thus of course also in Corollaries A.2.1 and A.2.2) can be weakened to the assumption that M has the direct sum property:

Definition A.2.1. *A measurable set M has the* direct sum property, *if there exists a family of pairwise disjoint sets M_α of finite measure with $M = \bigcup M_\alpha$, such that a function x is measurable on M, if and only if it is measurable on each M_α.*

The extension of Theorem A.2.1 is straightforward: Apply the σ-finite version of the theorem on each M_α to find a corresponding measurable function x_α, and then define $x = x_\alpha$ on M_α (this step needs the axiom of choice).

The direct sum property coincides with the direct sum property given in literature (like e.g. in [14]):

Proposition A.2.1. *A measurable set $M = \bigcup M_\alpha$ has the direct sum property with respect to the family $(M_\alpha)_{\alpha \in A}$ of pairwise disjoint sets of finite measure, if and only if for any set $E \subseteq M$ of finite measure there exists an at most countable number of indices α_n, such that $E \setminus \bigcup M_{\alpha_n}$ is a null set.*

Proof. First, let M have the direct sum property with respect to $(M_\alpha)_{\alpha \in A}$. We have to show that the above property holds true.
For this purpose observe that there exists a Banach space Y, and for any $\alpha \in A$ some $y_\alpha \in Y$, such that $|y_\alpha - y_\beta| \geq 1$ for $\alpha \neq \beta$. For example you might choose $Y = L_2(A)$ being the Hilbert space of square integrable functions $A \to \mathbb{R}$ with the counting measure on the index set A and let $y_\alpha = \chi_{\{\alpha\}}$ be the canonical orthonormal base of Y.
Now, define a mapping $x : M \to Y$ by $x(s) = y_\alpha$ for $s \in M_\alpha$. Since M has the direct sum property, x must be measurable. In particular, x must be essentially separable valued on each set E of finite measure, thus in view of Lemma A.1.1 even essentially countable valued on E. But by definition of x this means that E is contained in the union of countable many M_α (up to a null set).
Conversely, let M have the property of the statement with a family of sets M_α. Assume that $x : M \to Y$ is measurable on each M_α. We have to prove that x is the limit of simple functions on each set $E \subseteq M$ of finite measure, i.e. we have to prove that x is measurable on E. But by assumption there exist a countable number of indices α_n such that $E \setminus \bigcup M_{\alpha_n}$ is a null set. Thus it suffices to observe that each $x_n(s) = \chi_{M_{\alpha_n}}(s)x(s)$ is measurable with $x = \sum x_n$ a.e. on E. □

The direct sum property is a very weak property. It does not even imply the finite subset property:

Example A.2.1. Let S be the following measure space on $[0,1] \times [0,1]$: Call a set $E \subseteq S$ measurable, if each $E_t = \{s : (t,s) \in E\}$ is Lebesgue measurable. In this case let $\text{mes}E = \infty$ if there is an uncountable number of nonempty E_t, otherwise let $\text{mes}E = \sum_t \mu(E_t)$, μ being the Lebesgue measure. This measure space is discussed in [48, §10 Example 6], and it does not have the finite subset property and is not localizable (see [48, §34,§35]). However, $M = S$ has the direct sum property, since you may choose $M_\alpha = \{\alpha\} \times [0,1]$ for $\alpha \in [0,1]$.

This example shows also that, even if M has the direct sum property, it is not enough just to choose M_α of positive and finite measure such that M is 'exhausted': If you choose the family $M_\alpha = (\{\alpha\} \times [0,1]) \cup \{(1,\alpha)\}$ for $\alpha \in (0,1)$, $M_\alpha = (\{0\} \times [0,1]) \cup \{(1,0),(1,1)\}$ for $\alpha = 0$, the measurability of x on each M_α does not imply the measurability of x on M, since $x(1,\cdot)$ need not be Lebesgue measurable.

The following measure space may or may not have the direct sum property, depending on the system of axioms we use:

Example A.2.2. We consider the measure space discussed in [48, §10 Example 7] over $M = S = [0,1] \times [0,1]$. There a set $E \subseteq S$ is measurable if each $E_t = \{s : (t,s) \in E\}$ and each $E^s = \{t : (t,s) \in E\}$ is Lebesgue measurable. In this case define $\text{mes}E$ in the following way: If the number of nonempty E_t is at most countable, let $\text{mes}E = \sum_t \mu(E_t)$ (μ being the Lebesgue measure), and if the number of nonempty E^s is at most countable, let $\text{mes}E = \sum_s \mu(E^s)$. Otherwise let $\text{mes}E = \infty$.
Assume now that $M = \bigcup M_\alpha$ with M_α as in Definition A.2.1. Consider the sets $E_t = \{(t,s) : s \in [0,1]\}$ of finite measure. By Proposition A.2.1 for any t there is a countable number of set M_{α_n} such that $E_t \setminus \bigcup M_{\alpha_n}$ is a null set. In particular, if U denotes the union of all sets M_α such that $\{s : (t,s) \in M_\alpha\}$ is of positive measure for at least one t, then $U_t = \{s : (t,s) \in U\}$ has measure 1 for any t. Similarly, if V is the union of all sets M_α such that $\{t : (t,s) \in M_\alpha\}$ is of positive measure for at least one s we have that $V^s = \{t : (t,s) \in V\}$ has measure 1 for any s. Since all M_α have finite measure and are pairwise disjoint, also U and V are disjoint. But if U or V is Lebesgue measurable on $[0,1] \times [0,1]$ this is not possible by the Cavalieri principle.
Thus the assumption that M has the direct sum property allows us to construct a non Lebesgue measurable set on $[0,1] \times [0,1]$ without referring to the axiom of choice or the continuum hypothesis. But this is not possible in Solovay's model [44] (see also [23]). Hence (provided Solovay's model exists) we have shown that at least within Zermelo's set theory with the principle of dependent choices it is not possible to prove that M has the direct sum property.
But on the other hand, assuming the continuum hypothesis, Sierpinski showed [42] that there exists a set $D \subseteq [0,1] \times [0,1]$, such that each $D_t = \{s : (t,s) \in D\}$ is a null set and such that each $D^s = \{t : (t,s) \in D\}$ contains almost all points of $[0,1]$, see [40, 8.9(c)]. Now let $M_\alpha = (\{\alpha\} \times [0,1]) \setminus D$

for $\alpha \in [0,1]$, and $M_\alpha = ([0,1] \times \{\alpha - 2\}) \cap D$ for $\alpha \in [2,3]$. Then the M_α are pairwise disjoint and $\bigcup M_\alpha = M$. Moreover, if E has finite measure then either just a countable number of sets $E_t = \{s : (t,s) \in E\}$ or a countable number of sets $E^s = \{t : (t,s) \in E\}$ is nonempty. In the first case put $I = \{t : E_t \neq \emptyset\}$, in the second case put $I = \{s+2 : E^s \neq \emptyset\}$. In both cases I is at most countable, and $E \setminus \bigcup_{\alpha \in I} M_\alpha$ is a null set. Thus M has the direct sum property by Proposition A.2.1.

We now want to give an 'integral version' of Theorem A.2.1 and its corollaries. The σ-finite case is a straightforward application of the theorem. But surprisingly for integrals we may drop this assumption always. The reason for this is the following tricky

Lemma A.2.2. *Let S be some measure space, and $x, y : S \to [0,\infty]$ be measurable. Then there exists a sequence of measurable functions $0 \le x_n \le x$ with $x_n(s) < x(s)$ for $x(s) \neq 0$, such that*

$$\int_S x_n(s)y(s)ds \to \int_S x(s)y(s)ds.$$

Moreover, if $y : S \to [0,\infty)$ and $F = \operatorname{supp} x \cap \operatorname{supp} y$ has the finite subset property, you can choose x_n in such a way that each $x_n y$ is integrable.

Proof. For the first statement choose $x_n = \min\{n, \max\{x - n^{-1}, 0\}\}$ and use the monotone convergence theorem. Now, assume y is finite, and F has the finite subset property. Let $f(s) = x(s)y(s)$. It suffices to consider the case $\int_F f(s)ds = \infty$.

1. First, assume additionally that also x is finite everywhere. Let Γ be the set of all measurable E with $\int_E f(s)ds < \infty$. Choose a sequence $E_n \in \Gamma$ with

$$\int_{E_n} f(s)ds \to \sup_{E \in \Gamma} \int_E f(s)ds = \alpha.$$

Without loss of generality we may assume that $E_1 \subseteq E_2 \subseteq \ldots$ (otherwise replace E_n by $\bigcup_{k=1}^n E_k$). Let $U = \bigcup E_n$. By the monotone convergence theorem, we have

$$\int_{E_n} f(s)ds \to \int_U f(s)ds.$$

If $\alpha < \infty$, this implies $U \in \Gamma$. Thus $F \setminus U$ must have positive measure. But then for some k the set $\{s \in F \setminus U : f(s) \le k\}$ has positive measure, hence contains a subset D of positive but finite measure. But then also $U \cup D \in \Gamma$, which is by

$$\alpha = \int_U f(s)ds < \int_{U \cup D} f(s)ds$$

a contradiction to the definition of α. Thus $\alpha = \infty$. Now let $x_n(s) = \chi_{E_n}(s) \max\{x(s) - n^{-1}, 0\}$. Then we have by the monotone convergence theorem

$$\infty > \int_S x_n(s)y(s)ds \to \int_U f(s)ds = \alpha = \infty.$$

2. Now we drop the assumption that x is everywhere finite. But we may apply what we have shown for $w_n = \min\{x, n\}$ instead of x. Observing that $\int w_n(s)y(s)ds \to \infty$ by the monotone convergence theorem, we have that for any n there is some $0 \leq x_n \leq x$ with $x_n(s) < x(s)$ for $x(s) \neq 0$ with

$$n \leq \int_S x_n(s)y(s)ds < \infty.$$

Now the sequence $\max_{k \leq n} x_k$ is the sequence, we are looking for. □

Now we can prove the integral version of Theorem A.2.1 on *arbitrary* measure spaces:

Theorem A.2.2. *Let Y be a normed linear space, Z be a Banach space, and S be a measure space. Assume, $A : S \to \mathfrak{L}(Y, Z)$ is measurable. Let $y : S \to [0, \infty]$ be measurable, and M_y consist of all measurable functions $x : S \to Y$ with $|x(s)| = y(s)$ for $y(s) < \infty$. Then*

$$\sup_{x \in M_y} \int_S |A(s)x(s)| \, ds = \int_S |A(s)| \, y(s)ds,$$

where the case that a side is infinite is not excluded.

Proof. We prove the nontrivial inequality. Let $f(s) = |A(s)| \, y(s)$.
1. First, assume $T = \mathrm{supp} f$ has the finite subset property. Then by Lemma A.2.2 there exists a sequence $0 \leq y_n \leq y$ with $y_n < y$ on $\mathrm{supp} y$ and

$$\infty > \int_S |A(s)| \, y_n(s)ds \to \int_S f(s)ds.$$

Thus it remains to prove for any n that for $g(s) = |A(s)| \, y_n(s)$

$$\sup_{x \in M_y} \int_S |A(s)x(s)| \, ds \geq \int_S g(s)ds. \tag{A.2}$$

Observe that $\mathrm{supp} g$ is σ-finite, because it is the union of the sets $S_k = \{s : g(s) \geq k^{-1}\}$, which have by $\infty > \int g(s)ds \geq k^{-1}\mathrm{mes} S_k$ finite measure. Without loss of generality we may assume that $\mathrm{supp} A \subseteq \mathrm{supp} g$.
Fix n, and let $w(s) = y(s)$ for $y(s) < \infty$, $w(s) = y_n(s) + 1$ for $y(s) = \infty$. For $y(s) \neq 0$ let $z(s) = |A(s)| \, (1 - w(s)^{-1} y_n(s))$, otherwise $z(s) = |A(s)|$. Then $\mathrm{supp} z = \mathrm{supp} A$. If we apply Theorem A.2.1 for $W = \{y \in Y : |y| = 1\}$, we find that there exists a measurable $u : S \to W$ with

$$|A(s)u(s)| \geq |A(s)| - z(s) \qquad (s \in S).$$

Putting $x(s) = u(s)w(s)$, we have $x \in M_y$ and

$$|A(s)x(s)| \geq |A(s)|\, w(s) - z(s)w(s) = g(s) \qquad (s \in S).$$

Integrating this inequality yields (A.2).
2. Now, assume supp$f$ has not the finite subset property. This means, it contains a measurable set E of infinite measure, such that each measurable subset of E has either infinite measure or is a null set. But this implies that any function defined on E is measurable. Hence there exists some $x \in M_y$ with $A(s)x(s) \neq 0$ on E. Since for some k the set $W = \{s \in E : |A(s)x(s)| \geq k^{-1}\}$ has positive (whence infinite) measure, we thus have $\int_S |A(s)x(s)|\, ds \geq \int_W k^{-1} ds = \infty$. □

Again, the following corollaries continue to hold for complex Banach spaces:

Corollary A.2.3. *Let Y be some Banach space, Y^* its dual space, and S be some measure space. Let M be a set of measurable functions $S \to Y$ with the property that the conditions $x \in M$, $|z| = |x|$ for some measurable z imply $z \in M$. Then for any measurable $y : S \to Y^*$ we have*

$$\sup_{x \in M,\ yx \geq 0} \int_S y(s)x(s)ds = \sup_{x \in M} \int_S |y(s)|\, |x(s)|\, ds,$$

where the case that a side is infinite is not excluded.

Proof. Again, we just show the nontrivial inequality. Thus fix some $x \in M$. Applying Theorem A.2.2 for $A = y$ and $y = |x|$, we find a sequence of functions $z_n \in M$ with

$$\int_S |y(s)z_n(s)|\, ds \to \int_S |y(s)|\, |x(s)|\, ds.$$

Putting $x_n(s) = z_n(s)\mathrm{sgn}[\overline{y(s)z_n(s)}]$ (for complex $z \neq 0$ let sgn$z = z/|z|$), we have $x_n \in M, y(s)x_n(s) = |y(s)z_n(s)|$, hence $\int y(s)x_n(s)ds \to \int |y(s)|\, |x(s)|\, ds$.
□

For finite measure spaces, an analogous result may be found in [51, Lemma 3].

The 'dual' version of Corollary A.2.3 again uses the fact that the canonical embedding of Y into Y^{**} is norm-preserving (see Definition 2.1.8):

Corollary A.2.4. *Let Y be some Banach space with the bidual property, Y^* its dual space, and S be some measure space. Let M be a set of Y^*-valued measurable functions with the property that the conditions $y \in M$, $|z| = |y|$ for some measurable z imply $z \in M$. Then for any measurable $x : S \to Y$ we have*

$$\sup_{y \in M,\ yx \geq 0} \int_S y(s)x(s)ds = \sup_{y \in M} \int_S |y(s)|\, |x(s)|\, ds,$$

where the case that a side is infinite is not excluded.

Proof. Let $Z^* = Y^{**}$ be the dual space of $Z = Y^*$. Let $i : Y \to Z^*$ be the canonical embedding. If we define $z : S \to Z^*$ by $z = i \circ x$, i.e. $z(s)l = l(x(s))$, z is measurable, because i is continuous. If we apply Corollary A.2.3 for Z instead of Y, and for z instead of y, we find

$$\sup_{y \in M,\ zy \geq 0} \int_S z(s)y(s)ds = \sup_{y \in M} \int_S |z(s)|\,|y(s)|\,ds.$$

This is the stated equality, since $z(s)y(s) = y(s)x(s)$ and $|z(s)| = |x(s)|$, because Y has the bidual property. □

A.3 A Generalization of a Theorem of Luxemburg-Gribanov

At first, we recall a theorem of Luxemburg [31] and Gribanov [21] (see [49, Theorem 99.2 and Corollary 99.3]). In the following, let T and S be σ-finite measure spaces.

Theorem A.3.1. *Let B be some set of measurable nonnegative functions over S, and x be measurable and nonnegative on $T \times S$. Then there exists a countable set $A \subseteq B$, such that for almost all t*

$$\sup_{u \in B} \int_S x(t,s)u(s)ds = \sup_{u \in A} \int_S x(t,s)u(s)ds.$$

In particular, $t \mapsto \sup_{u \in B} \int_S x(t,s)u(s)ds$ is measurable.

Proof. 1. We first show that we additionally may assume that there exists some set of finite measure $E \subseteq S$, such that $\operatorname{supp} x \subseteq T \times S$, and that all $u \in B$ and x are uniformly bounded by some $C > 0$.
Indeed, assume the statement has been shown for this case. Let $S = \bigcup S_n$, where $S_1 \subseteq S_2 \subseteq \ldots$ have finite measure, and put $x_n = P_{T \times S_n} \min(x, n)$, and $B_n = \{\min(u, n) : u \in B\}$. Since for any $u \in B$ by the monotone convergence theorem (for almost all t)

$$\int_S x_n(t,s) \min(u(s), n) ds \to \int_S x(t,s)u(s)ds,$$

we have

$$\sup_n \sup_{u \in B_n} \int_S x_n(t,s)u(s)ds = \sup_{u \in B} \int_S x(t,s)u(s)ds.$$

By assumption, each B_n contains a countable A_n satisfying

$$\sup_{u \in B_n} \int_S x_n(t,s)u(s)ds = \sup_{u \in A_n} \int_S x_n(t,s)u(s)ds.$$

Putting $A^* = \bigcup A_n$, we thus have by taking the supremum over all n:

$$\sup_{u \in A^*} \int_S x(t,s)u(s)ds = \sup_{u \in B} \int_S x(t,s)u(s)ds.$$

By construction, for each $u^* \in A^*$ there exists a function $u \in B$ with $u \geq u^*$. Thus the statement is true for A being the set of all that u.
2. Now, assume that x is elementary, i.e. (Lemma 4.3.2)

$$x(t,s) = \sum_{k=1}^{m} \chi_{T_k}(t) x_k(s),$$

with simple functions $x_k \geq 0$ and measurable pairwise disjoint $T_k \subseteq T$, $\bigcup T_k = T$. For any k and any n there exists some $u_{kn} \in B$ satisfying

$$\sup_{u \in B} \int_S x_k(s)u(s)ds \leq \int_S x_k(s)u_{kn}(s)ds + n^{-1}.$$

Then for almost all $t \in T_k$, we have by $x(t,s) = x_k(s)$ that

$$\sup_{u \in B} \int_S x(t,s)u(s)ds = \sup_n \int_S x(t,s)u_{kn}(s)ds,$$

i.e. the statement is true for A being the set of all u_{kn}.
3. We now put the additional assumptions of 1. on B and x. Let y_n be elementary, converging to x a.e. (Lemma 4.3.3). Then $x_n = P_{T \times E} \min(C, |y_n|) \geq 0$ converges to x, but additionally satisfies $0 \leq x_n \leq C$ and $\mathrm{supp} x_n \subseteq T \times E$. Thus Lebesgue's dominated convergence theorem implies (for almost all t)

$$\begin{aligned} \lim_{n\to\infty} \sup_{u \in B} \left| \int_S x(t,s)u(s)ds - \int_S x_n(t,s)u(s)ds \right| \\ \leq \lim_{n\to\infty} \int_S |x(t,s) - x_n(t,s)|\, C\, ds = 0. \end{aligned} \tag{A.3}$$

By 2., for any n, there exists some countable $A_n \subseteq B$ satisfying

$$\sup_{u \in B} \int_S x_n(t,s)u(s)ds \leq \sup_{u \in A_n} \int_S x_n(t,s)u(s)ds. \tag{A.4}$$

Now, given some $\varepsilon > 0$, by (A.3) there exists some n with

$$\int_S x_n(t,s)u(s)ds - \varepsilon \leq \int_S x(t,s)u(s)ds \leq \int_S x_n(t,s)u(s)ds + \varepsilon \qquad (u \in B).$$

Thus, putting $A = \bigcup A_n$, we have by (A.4) for that n (and almost all t)

$$\begin{aligned} \sup_{u \in B} \int_S x(t,s)u(s)ds &\leq \sup_{u \in B} \int_S x_n(t,s)u(s)ds + \varepsilon \\ &\leq \sup_{u \in A} \int_S x_n(t,s)u(s)ds + \varepsilon \leq \sup_{u \in A} \int_S x(t,s)u(s)ds + 2\varepsilon. \end{aligned}$$

This shows the statement, since ε was arbitrary. □

We need the generalization, that for any error-function y, one can choose a measurable majorizing function:

Theorem A.3.2. *Let B be some set of measurable nonnegative functions over T, and x be measurable and nonnegative on $T \times S$. Assume, y is nonnegative and measurable with $y(t) > 0$ for $x(t,\cdot) \neq 0$. Then there exists a nonnegative measurable function u on $T \times S$, such that for almost all t we have $u(t,\cdot) \in B$ and*

$$\int_S x(t,s)u(t,s)ds \geq \sup_{u \in B} \int_S x(t,s)u(s)ds - y(t).$$

Moreover, the function $t \mapsto u(t,\cdot)$ is just countable-valued.

Proof. We write $Fu(t) = \int_S x(t,s)u(s)ds$ and $F(t) = \sup_{u \in B} Fu(t)$ for short. By Theorem A.3.1, $F(t)$ is measurable, and there exists a sequence $u_n \in B$ with

$$F(t) = \sup_n Fu_n(t). \tag{A.5}$$

Now, let

$$E_n = \{t \in T : Fu_n(t) \geq F(t) - y(t)\}.$$

Then E_n belongs to the Lebesgue extension of the measure, and $T \setminus \bigcup E_n$ is a null set by (A.5). Now put

$$D_n = E_n \setminus \bigcup_{k<n} E_k.$$

Then D_n are measurable, pairwise disjoint, and satisfy $\bigcup D_n = \bigcup E_n$. Thus, the function $u(t,s) = \sum \chi_{D_n \times S}(t,s)u_n(s)$ is measurable on $T \times S$, and for almost all $t \in T$, we have that $t \in D_n$ for some n, and thus by $D_n \subseteq E_n$

$$Fu(t,\cdot)(t) = Fu_n(t) \geq F(t) - y(t).$$

Whence, u is the desired function. □

The previous theorems also have obvious countable generalizations:

Theorem A.3.3. *Let $B(\cdot)$ be a family of sets of nonnegative measurable functions over S, which is countable measurable on T. Let x be measurable and nonnegative on $T \times S$. Then there exists a countable measurable family $A(t) \subseteq B(t)$ of countable sets, such that for almost all t:*

$$\sup_{u \in B(t)} \int_S x(t,s)u(s)ds = \sup_{u \in A(t)} \int_S x(t,s)u(s)ds.$$

In particular, $t \mapsto \sup_{u \in B(t)} \int_S x(t,s)u(s)ds$ is measurable.

Theorem A.3.4. *Let $B(\cdot)$ be a family of sets of nonnegative measurable functions over S, which is countable measurable on T. Let x be measurable and nonnegative on $T \times S$. Assume, y is nonnegative and measurable with $y(t) > 0$ for $x(t,\cdot) \neq 0$. Then there exists a nonnegative measurable function u on $T \times S$, such that for almost all t we have $u(t,\cdot) \in B(t)$ and*

$$\int_S x(t,s)u(t,s)ds \geq \sup_{u \in B(t)} \int_S x(t,s)u(s)ds - y(t).$$

Moreover, the function $t \mapsto u(t,\cdot)$ is just countable-valued.

For the proofs, observe that there exist a countable number of sets B_n and measurable pairwise disjoint $T_n \subseteq T$ with $\bigcup T_n = T$, such that $B(t) = B_n$ for almost all $t \in T_n$. Thus, it suffices to apply Theorem A.3.1 on each T_n instead of T to find countable $A_n \subseteq B_n$, and to put $A(t) = A_n$ for $t \in T_n$. This shows Theorem A.3.3. To prove Theorem A.3.4, similarly apply Theorem A.3.2 for each T_n instead of T to find a function u_n, and then put $u(t,s) = u_n(t,s)$ for $t \in T_n$.

References

1. Alt, H.W. (1992): Lineare Funktionalanalysis, 2nd ed., Springer, Berlin Heidelberg New York.
2. Amemiya, I. (1953): A generalization of Riesz-Fischer's theorem, J. Math. Soc. Japan **5**, 353–354.
3. Appell, J., Zabrejko, P.P. (1989): Continuity properties of the superposition operator, J. Austral. Math. Soc. Ser. A **47**, 186–210.
4. Appell, J., Zabrejko, P.P. (1990): Nonlinear superposition operators, Cambridge Univ. Press, Cambridge.
5. Appell, J., Diallo, O.W., Zabrejko, P.P. (1988): On linear integro-differential equations of Barbashin type in spaces of continuous and measurable functions, J. Integr. Equ. Appl. **1**, (2), 227–247.
6. Benedek, A., Panzone, R. (1961): The spaces L^P, with mixed norm, Duke Math. J. **28**, 301–324.
7. Bochner, S., Taylor, A.E. (1938): Linear functionals on certain spaces of abstractly-valued functions, Ann. of Math. (2) **39**, 913–944.
8. Burgess, D.C.J. (1954): Abstract moment problems with applications to the l^p and L^p spaces, Proc. London Math. Soc. (3) **4**, 107–128.
9. Chen, C.-J. (1995): On a generalized integro-differential equation of Barbashin type, Z. Anal. Anw. **14**, (4), 899–912.
10. Chen, C.-J., Väth, M. (1997): On the $\mathfrak{L}$-characteristic of the superposition operator in Lebesgue spaces with mixed norm, Z. Anal. Anw. **16**, (2), 377–386.
11. Daleckiĭ, Ju.L., Krein, M.G. (1974): Stability of solutions of differential equations in Banach space (in Russian), Transl. Math. Monographs, vol.43, Amer. Math. Soc., Providence, R. I.
12. Deimling, K. (1985): Nonlinear functional analysis, Springer, Berlin Heidelberg.
13. Diallo, O.W. (1988): Methods of functional analysis in the theory of linear integro-differential equations of Barbashin type (in Russian), Kand. dissertation, University of Minsk, Minsk.
14. Dinculeanu, N. (1967): Vector measures, Pergamon Press, Oxford / VEB Deutscher Verlag der Wissenschaften, Berlin.
15. Dunford, N., Schwartz, J. (1966): Linear operators I, 3rd ed., Int. Publ., New York.
16. Ermolova, E.A. (1995): Ljapunov-Bohl-Exponent und Greensche Funktion für eine Klasse von Integro-Differentialgleichungen, Z. Anal. Anw. **14**, (4), 881–898.
17. Garnir, H.G. (1974): Solovay's axiom and functional analysis, in Garnir et al. [19], 189–204.

18. Garnir, H.G., De Wilde, M., Schmets, J. (1968): Analyse fonctionnelle, Théorie constructive des espaces linéaires à semi-normes, tome I: Théorie générale, Birkhäuser, Basel Stuttgart.
19. Garnir, H.G., Unni, K.R., Williamson, J.H. (eds.) (1974): Proc. Madras conference functional analysis 1973, Lect. Notes Math., no. 399, Springer, Berlin.
20. Gel'fand, I.M. (1938): Abstrakte Funktionen und lineare Operatoren, Mat. Sbornik **4**, 235–286.
21. Gribanov, Ya.I. (1970): The measurability of a function (in Russian), Izv. Vyssh. Uchebn. Zaved. Mat. **3**, 22–26.
22. Hille, E., Phillips, R.S. (1957): Functional analysis and semi-groups, Amer. Math. Soc. Coll. Publ., Providence, R. I.
23. Jech, T.J. (1971): Lectures in set theory with particular emphasis on the method of forcing, Lect. Notes Math., no. 217, Springer, Berlin.
24. Jech, T.J. (1973): The axiom of choice, Studies in Logic and the Foundations of mathematics, vol.75, North-Holland Publ. Company, Amsterdam.
25. Jörgens, K. (1970): Lineare Integraloperatoren, Teubner, Stuttgart.
26. Kalitvin, A.S., Zabrejko, P.P. (1991): On the theory of partial integral operators, J. Integr. Equ. Appl. **3**, (3), 351–382.
27. Krasnoselskiĭ, M.A. (1951): The continuity of a certain operator, Dokl. Akad. Nauk SSSR **77**, (2), 185–188.
28. Krasnoselskiĭ, M.A., Ladyzhenskiĭ, L.A. (1954): Conditions for the complete continuity of the P. S. Uryson operator (in Russian), Trudy Moskov. Mat. Obshch. **3**, 307–320.
29. Krasnoselskiĭ, M.A., Zabrejko, P.P., Pustylnik, E.I., Sobolevskiĭ, P.E. (1976): Integral operators in spaces of summable functions (in Russian), Noordhoff, Leyden.
30. Kufner, A., Oldrich, J., Fučík, S. (1977): Function spaces, Noordhoff, Leyden; Prague, Academia.
31. Luxemburg, W.A.J. (1958): On the measurability of a function which occurs in a paper by A. C. Zaanen, Proc. Netherl. Acad. Sci. (A) **61**, 259–265, (Indag. Math. **20**, 259-265).
32. Luxemburg, W.A.J. (1963): Addendum to "On the measurability of a function which occurs in a paper by A. C. Zaanen", Proc. Roy. Acad. Sci. **A61**, 587–590.
33. Luxemburg, W.A.J. (ed.) (1969): (Intern. symp. on the) Applications of model theory to algebra, analysis, and probability (1967), Toronto, Holt, Rinehart and Winston.
34. Luxemburg, W.A.J. (1969): Reduced powers of the real number system and equivalents of the Hahn-Banach extension theorem, in (Intern. Symp. on the) Applications of Model Theory to Algebra, Analysis, and Probability (1967) [33], 123–137.
35. Luxemburg, W.A.J. (1973): What is nonstandard analysis?, Amer. Math. Monthly **80**, (6 part II), 38–67.
36. Mullins, C.W. (1974): Linear functionals on vector valued Köthe spaces, in Garnir et al. [19], 380–381.
37. Pincus, D. (1973): The strength of the Hahn-Banach theorem, Victoria Symposium on Nonstandard Analysis (Univ. of Victoria 1972) (Dold, A., Eckmann, B., eds.), Lect. Notes Math., no. 369, Springer, Berlin, 203–248.
38. Povolotskiĭ, A.I., Kalitvin, A.S. (1991): Nonlinear partial integral operators (in Russian), Leningrad.
39. Rao, M.M., Ren, Z.D. (1991): Theory of Orlicz spaces, M. Dekker, New York.
40. Rudin, W. (1987): Real and complex analysis, 3rd ed., McGraw-Hill, Singapore.
41. Schaefer, H.H. (1974): Banach lattices and positive operators, Springer, Berlin Heidelberg New York.

42. Sierpiński, W. (1920): Sur les rapports entre l'existence des intégrales, Fund. Math. **1**, 142–147.
43. Sierpiński, W. (1938): Fonctions additives non complètement additives et fonctions non mesurables, Fund. Math. **30**, 96–99.
44. Solovay, R.M. (1970): A model of set-theory in which every set of reals is Lebesgue measurable, Ann. of Math. (2) **92**, 1–56.
45. Taylor, R.F. (1969): On some properties of bounded internal functions, in Luxemburg [33], 167–170.
46. Väth, M. (1993): Lineare und nichtlineare Volterra-Gleichungen, Diplomarbeit, University of Würzburg, Würzburg.
47. Wright, J.D.M. (1977): Functional analysis for the practical man, Functional Analysis: Surveys and Recent Results (Proceedings of the Conference on Functional Analysis, Paderborn, Germany 1976) (Amsterdam) (Bierstedt, K.-D., Fuchssteiner, B., eds.), North-Holland Math. Stud. 27 / Notas de Matemática 63, North-Holland, 283–290.
48. Zaanen, A.C. (1967): Integration, North-Holland Publ. Company, Amsterdam.
49. Zaanen, A.C. (1983): Riesz spaces, vol.II, North-Holland Publ. Company, Amsterdam New York Oxford.
50. Zabrejko, P.P. (1974): Ideal spaces of functions I (in Russian), Vestnik Jarosl. Univ., 12–52.
51. Zabrejko, P.P., Obradovich, P. (1968): On the theory of Banach vector function spaces (in Russian), Voronezh. Gos. Univ. Trudy Sem. Funk. Anal., 431–440.

Index